System Reliability
Management

Advanced Research in Reliability and System Assurance Engineering

Series Editor:
Mangey Ram
Professor, Assistant Dean (International Affairs), Department of Mathematics, Graphic Era Deemed to be University, Dehradun, India

Reliability Engineering: Theory and Applications
Edited by Ilia Vonta and Mangey Ram

Modeling and Simulation-Based Analysis in Reliability Engineering
Edited by Mangey Ram

System Reliability Management: Solutions and Technologies
Edited by Adarsh Anand and Mangey Ram

For more information about this series, please visit: https://www.crcpress.com/Reliability-Engineering-Theory-and-Applications/Vonta-Ram/p/book/9780815355175

System Reliability
Management
Solutions and Technologies

Edited by
Adarsh Anand and Mangey Ram

CRC Press
Taylor & Francis Group
Boca Raton London New York

CRC Press is an imprint of the
Taylor & Francis Group, an **informa** business

CRC Press
Taylor & Francis Group
6000 Broken Sound Parkway NW, Suite 300
Boca Raton, FL 33487-2742

First issued in paperback 2020

© 2019 by Taylor & Francis Group, LLC
CRC Press is an imprint of Taylor & Francis Group, an Informa business

No claim to original U.S. Government works

ISBN-13: 978-0-8153-6072-8 (hbk)
ISBN-13: 978-0-367-78078-4 (pbk)

Library of Congress Cataloging-in-Publication Data

Names: Anand, Adarsh, author. | Ram, Mangey, author.
Title: System reliability management : solutions and technologies / Adarsh Anand & Mangey Ram.
Description: Boca Raton : Taylor & Francis, a CRC title, part of the Taylor & Francis imprint, a member of the Taylor & Francis Group, the academic division of T&F Informa, plc, [2019] | Series: Advanced research in reliability and system assurance engineering | Includes bibliographical references and index.
Identifiers: LCCN 2018021249 | ISBN 9780815360728 (hardback : acid-free paper) | ISBN 9781351117661 (ebook)
Subjects: LCSH: Reliability (Engineering)
Classification: LCC TA169 .A624 2019 | DDC 620/.00452—dc23
LC record available at https://lccn.loc.gov/2018021249

Visit the Taylor & Francis Web site at
http://www.taylorandfrancis.com

and the CRC Press Web site at
http://www.crcpress.com

Dr. Adarsh Anand would like to dedicate this edited book to his parents;

his other half, Dr. Deepti Aggrawal; and daughter, Ahana Anand.

Prof. Mangey Ram would like to dedicate this edited

book to his family members and friends.

Contents

Preface...ix

Acknowledgments..xi

Authors...xiii

List of Contributors xv

1. AI Approach to Fault Big Data Analysis and Reliability
   Assessment for Open-Source Software 1
   Yoshinobu Tamura and Shigeru Yamada

2. Modeling Software Fault Removal and Vulnerability Detection
   and Related Patch Release Policy.. 19
   Adarsh Anand, Priyanka Gupta, Yury Klochkov, and V. S. S Yadavalli

3. System Reliability Optimization in a Fuzzy Environment via
   Hybridized GA–PSO 35
   Laxminarayan Sahoo

4. Optimal Software Testing Effort Expending Problems......................51
   Shinji Inoue and Shigeru Yamada

5. Revisiting Error Generation and Stochastic Differential
   Equation-Based Software Reliability Growth Models 65
   Adarsh Anand, Deepika, A. K. Verma, and Mangey Ram

6. Repairable System Modeling Using Power Law Process 79
   K. Muralidharan

7. Reliability and Safety Management of Engineering Systems
   through the Prism of "Black Swan" Theory 103
   Iosif Aronov and Ljubisa Papic

8. Reliable Recommender System Using Improved Collaborative
   Filtering Technique.. 113
   Rahul Katarya

9. Categorization of Vulnerabilities in a Software................................. 121
   Navneet Bhatt, Adarsh Anand, Deepti Aggrawal, and Omar H. Alhazmi

10. **Reliable Predictions of Peak Electricity Demand and Reliability of Power System Management**.. 137
Caston Sigauke, Santosh Kumar, Norman Maswanganyi, and Edmore Ranganai

11. **Real-Time Measurement and Evaluation as System Reliability Driver** 161
Mario José Diván and María Laura Sánchez Reynoso

12. **Optimizing Price, Release, and Testing Stop Time Decisions of a Software Product** 189
A. K. Shrivastava, Subhrata Das, Adarsh Anand, and Ompal Singh

13. **Software Reliability: A Quantitative Approach** 205
Yashwant K. Malaiya

14. **Modeling and Analysis of Fault Dependency in Software Reliability Growth Modeling** 237
V. B. Singh

Index 253

Preface

Because a multifaceted aspect of research allows us to answer critical queries and address broad issues, studying associative and collaborative fields is the need of the hour. Hence, the aim of this book is to provide a platform for researchers, industries, managers, and government policy makers to cooperate and collaborate among themselves to improve the reliability prediction and maintenance procedures, and to comprehend all the enterprise operations, maintenance, reliability, and engineering so as to manage their resources and improve their utilization for the benefit of society with organizational goals.

Due to the criticality and complexity of modern systems, there has arisen an ever-increasing attention to look for products with higher reliability at a reasonable cost. The greatest problem being faced by industries nowadays is understanding how to assess the reliability (both hardware and software) quantitatively. This book aims to present the state-of-the art of system reliability in theory and practice. It is an edited book based on the contributions made by researchers working in the reliability management field. Emphasis has been given on the original and qualitative work relevant to the theme, with particular importance given to system reliability management.

This edited issue of *Advance Research in Reliability and System Assurance Engineering* on *System Reliability Management: Solutions and Technologies* includes invited papers appropriate to the theme and the complex solution approaches to handle the research challenges in the related domain. The topics covered are organized as follows:

Chapter 1 discusses the concept of artificial intelligence in reliability assessment for open-source software. It explains the usage of neural network and deep learning approach in fault quantification.

Chapter 2 discusses the different treatments given to faults and vulnerabilities existing in a software and presents an optimal patch release policy.

Chapter 3 describes the utility of hybrid genetic algorithm and particle swarm optimization in determining the system reliability in a fuzzy environment.

Chapter 4 provides an optimization model consisting of software testing and maintenance cost. Optimal testing effort expenditure is evaluated, and related policies based on cost and simultaneous cost and reliability criteria are discussed.

Chapter 5 explains the reliability modeling framework based on the stochastic differential equation that is capable of dealing with all the available forms of error generations.

Chapter 6 explores the utility of power law process in modeling the repairable systems such as machines, industrial plants, and software.

Chapter 7 discusses the safety of engineering systems through the prism of black swan theory.

Chapter 8 deals with the reliable recommender system using improved typicality-based collaborative filtering technique.

Chapter 9 describes a detailed outlook toward categorization of vulnerabilities present in a software system.

Chapter 10 presents an analysis of system reliability and applicability of regression models for determining peak demand of electricity.

Chapter 11 demonstrates an idea of data and information in the context of the heterogeneous sensor networks followed by the updated Context-Indicator, Concept Model, Attribute, Metric and Indicator (C-INCAMI) Measurement and Evaluation framework.

Chapter 12 deals with a model that determines the optimal release and testing stop time that considers the impact of expected software sales, its pricing, and the software failure reported by customers during the warranty period.

Chapter 13 provides a detailed literature review in the field of software reliability assessment. In this chapter, various software system reliability measures are considered and their quantification is thoroughly explained.

Chapter 14 presents a detailed review of modeling and analysis of fault dependency in software reliability growth models.

The book shall be a valuable tool for practitioners and managers in system reliability engineering.

The editors are grateful to various authors across the globe who have contributed to this editorial work. All the reviewers who have helped through their comments and suggestions in improving the quality of the chapters deserve significant praise for their assistance. Gratitude also goes to Mr Navneet Bhatt, research scholar in the department of operational research, who has helped in carrying out the entire process in a streamlined manner.

Adarsh Anand
University of Delhi, India

Mangey Ram
Graphic Era Deemed to be University, India

Acknowledgments

The editors acknowledge CRC Press for providing this opportunity and professional support. They thank all the chapter contributors and reviewers for their availability for this work.

Authors

Dr. Adarsh Anand completed his doctorate from the Department of Operational Research, University of Delhi, India, and is currently associated with the same university as an Assistant Professor. Prior to joining as an Assistant Professor, Dr. Anand worked as University Teaching Assistant in the Department of Operational Research. He qualified the CSIR-NET (National Eligibility Test) for lecturer-ship in June 2010. His research area includes modeling of innovation adoption in marketing and software reliability assessment. He has publications in journals of international and national repute. He is a lifetime member of the Society for Reliability Engineering, Quality, and Operations Management. He has been conferred with the International Research Excellence Award for Contributions for "Young Promising Researcher in the Field of Innovation Diffusion Modeling in Marketing" during the 7th International Conference on Quality, Reliability, Infocom Technology, and Business Operations (ICQRITBO 2015). In 2014, he was conferred with the Amity Award for "Best Academician Research Paper Presentation" on the occasion of International Business School Conference (IBSCON 2014) by Amity International Business School, Amity University, Uttar Pradesh, India. He has also been awarded "Young Promising Researcher in the Field of Software Reliability and Technology Management" by the Society for Reliability Engineering, Quality, and Operations Management (SREQOM 2012).

Dr. Mangey Ram received his PhD degree with major in mathematics and minor in computer science from G. B. Pant University of Agriculture and Technology, Pantnagar, India. He has been a faculty member for around 10 years and has taught several core courses in pure and applied mathematics at undergraduate, postgraduate, and doctorate levels. He is currently a professor at Graphic Era Deemed to be University, Dehradun, India. Before joining Graphic Era, he was a deputy manager (probationary officer) with Syndicate Bank for a short period. He is Editor-in-Chief of *International Journal of Mathematical, Engineering, and Management Sciences* and the guest editor and member of the editorial board of various journals. He is a regular reviewer for international journals, including IEEE, Elsevier, Springer, Emerald, John Wiley, Taylor & Francis Group, and many other publishers. He has published 131 research publications in IEEE, Taylor & Francis, Springer, Elsevier, Emerald, World Scientific, and many other national and international journals of repute and also presented his works at national and international conferences. His fields of research include reliability theory and applied mathematics. Dr. Ram is a senior member of the IEEE; a life member of the Operational Research Society of India, the Society for

Reliability Engineering, Quality and Operations Management in India, and the Indian Society of Industrial and Applied Mathematics; and a member of the International Association of Engineers in Hong Kong and Emerald Literati Network in the UK. He has been a member of the organizing committee of a number of international and national conferences, seminars, and workshops. He has been conferred with the "Young Scientist Award" by the Uttarakhand State Council for Science and Technology, Dehradun in 2009. He has received the "Best Faculty Award" in 2011 and recently the "Research Excellence Award" in 2015 for his significant contribution in academics and research at Graphic Era.

List of Contributors

Deepti Aggrawal
Jaypee Institute of Information
    Technology,
Noida, UP, India

Omar H. Alhazmi
Taibah University
Medina, Saudi Arabia

Adarsh Anand
University of Delhi
New Delhi, India

Iosif Aronov
Department of Technical Barriers
    Analysis
Research Center "International
    Trade and Integration"
Moscow, Russia

Navneet Bhatt
University of Delhi
New Delhi, India

Subhrata Das
University of Delhi
New Delhi, India

Deepika
University of Delhi
New Delhi, India

Mario José Diván
National University of La Pampa
Santa Rosa, Argentina

Priyanka Gupta
University of Delhi
New Delhi, India

Shinji Inoue
Kansai University
Osaka, Japan

Rahul Katarya
Department of Computer Science &
    Engineering
Delhi Technological University
New Delhi, India

Yury klochkov
St. Petersburg Polytechnic
    University
St. Petersburg, Russia

Santosh Kumar
Department of Mathematics and
    Statistics
University of Melbourne
Melbourne, Australia

Yashwant K. Malaiya
Computer Science Department
Colorado State University
Fort Collins, Colorado

Norman Maswanganyi
Department of Statistics and
    Operations Research
University of Limpopo
Polokwane, South Africa

K. Muralidharan
The Maharaja Sayajirao University
    of Baroda
Vadodara, India

Ljubisa Papic
DQM Research Center
Prijevor, Serbia

Mangey Ram
Graphic Era Deemed to be
    University
Dehradun, Uttarakhand, India

Edmore Ranganai
Department of Statistics
University of South Africa
Pretoria, South Africa

María Laura Sánchez Reynoso
National University of La Pampa
Santa Rosa, Argentina

Laxminarayan Sahoo
Raniganj Girls' College
Burdwan, West Bengal, India

A.K.Shrivastava
Fortune Institute of International
    Business
New Delhi, India

Caston Sigauke
Department of Statistics
University of Venda
Thohoyandou, South Africa

Ompal Singh
University of Delhi
New Delhi, India

V.B. Singh
Department of Computer Science
Delhi College of Arts & Commerce,
    University of Delhi
New Delhi, India

Yoshinobu Tamura
Tokyo City University
Tokyo, Japan

A. K. Verma
Western Norway University of
    Applied Science
Haugesund, Norway

V.S.S. Yadavalli
University of Pretoria,
South Africa

Shigeru Yamada
Tottori University
Tottori, Japan

1

AI Approach to Fault Big Data Analysis and Reliability Assessment for Open-Source Software

Yoshinobu Tamura

Tokyo City University

Shigeru Yamada

Tottori University

CONTENTS

1.1 Introduction 1
1.2 Open Source Software 2
1.3 Fault Data on BTS of OSS 3
1.4 AI Approach 4
    1.4.1 NN Approach 4
    1.4.2 DL Approach 6
    1.4.3 Time Series Analysis 7
1.5 Numerical Examples 9
    1.5.1 Component Identification for Software Failure 9
    1.5.2 Reliability Analysis Based on Time Series 11
1.6 Conclusion 14
Acknowledgments 16
References 16

1.1 Introduction

Various software reliability models have been developed by related researchers in the past [1–3]. At present, artificial intelligence (AI) is used for assisting the achievement of unmanned systems in various research areas. In particular, deep learning (DL) is the hot topic currently focused on AI. In the research area of software engineering, open source software (OSS) is considered as the useful software maintenance and development paradigm in software development, management, and maintenance because of standardization, reduction in cost, short delivery, and so on [4]. In the development cycle of OSS, the bug tracking system (BTS) helps the OSS developers and managers to improve the quality of OSS. The amount of data in the BTS thus becomes large in various OSS projects.

This chapter discusses several AI approaches to fault big data analysis for OSS projects. Moreover, this chapter compares the conventional neural network (NN) approach with the proposed DL approach. This is followed by various numerical examples based on the fault big data in practical OSS projects by using the proposed methods of AI. Finally, this chapter shows that the proposed AI methods can improve the quality of OSS managed and developed by the practical project.

1.2 Open Source Software

Various OSSs have been managed, developed, and used under various practical OSS projects. Poor handling of reliability and quality problem in terms of OSS is a main concern, because the management and development cycles of OSS have no clear testing phase. The BTSs are especially used and managed in various OSS projects. The large-scale fault data are registered on these BTSs. The OSS reliability and quality can be improved significantly if the software users, managers, and developers can make an efficient and effective use of the fault big data sets on the BTSs.

In recent history, various software reliability models [1–3] have been used and applied to assess and improve the quality/reliability of OSS and test their progress during the software development process. Also, various OSS reliability growth models for reliability and quality assessment of OSS have been developed and proposed [4]. However, it is difficult for the OSS project managers and developers to select the optimal model for the practical software project. For example, they can assess the software reliability for the actual data sets by using the evaluation criteria of model. Therefore, it is very difficult for software developers and managers to analyze and assess the reliability and quality of OSS by using conventional software reliability growth models. Moreover, they will need to change the raw data on BTS to fault count data. It is efficient to compare the past methods based on conventional models if the OSS developers and managers can utilize all raw data sets of BTS.

This chapter focuses on the fault identification method that determines the severity level of the fault. Also, the OSS quality/reliability assessment method based on DL is discussed by using the mean time between software failures. Various numerical examples of OSS reliability analysis for the fault big data in the practical OSS projects have also been shown. Furthermore, this chapter compares the analysis method based on the DL with that based on NN by back propagation learning.

1.3 Fault Data on BTS of OSS

In general, the fault big data of various OSSs have been analyzed and managed on the software system for management and development support known as the BTS. The proprietary software such as BugLister and the OSS such as Bugzilla are known as the BTS. The data contents registered as the fault big data in OSS are as follows.

The Data Contents for OSS

- Bug ID number for each bug of OSS
- OSS product name for each bug
- OSS component for each bug
- Nickname of OSS assignee
- OSS status for each bug
- OSS resolution for each bug
- Summary for each bug of OSS
- OSS changed date and time
- OSS alias
- Assignee real name for each bug of OSS
- Hardware in usage of OSS
- Keywords for each bug
- Number of real comments for each bug
- OSS opened date and time
- Operating system (OS) for each bug
- Bug priority
- Nickname of OSS reporter
- Reporter real name for each bug of OSS
- Bug severity for each bug
- Bug tags for each bug
- OSS target milestone
- URL for each bug
- OSS version for each bug
- Votes for each bug

It will be helpful for the OSS developers, users, and OSS project managers if the software project managers can analyze and assess the abovementioned fault contents. There are several useful software tools to visualize the fault big data for specified organizations. However, there is no useful tool to assess the fault big data registered on the BTS.

For example, there are useful software assessment tools in terms of fault such as Redmine, Jira, and YouTrack. Jira and YouTrack are the proprietary software. Moreover, Redmine is known as the OSS. Also, TeamCity and Jenkins are well known as the continuous integration server. However, most of these support tools for software management cannot automatically evaluate and assess the quality based on the software fault data. This chapter discusses the OSS fault identification method based on DL.

1.4 AI Approach

AI is currently used in almost every field. For example, the DL approach by using big data is helpful in the medical imaging system of medical field, industrial business strategy, weather data analysis, risk assessment, visual analysis, and log analysis. On the other hand, several traditional approaches based on the stochastic models are used in the areas of software testing, project management, and software development. Recently, various data in terms of software testing, project management, and software development are registered via a computer network in the world. The AI technology will lead to a new approach for software project management by using these big data in terms of software development, testing, and management. Several AI approaches such as DL and NN for OSS projects are discussed in this chapter.

1.4.1 NN Approach

The framework of the NNs in this chapter is shown in Figure 1.1. Let $w_{ij}^1 (i = 1, 2, ..., I; j = 1, 2, ..., J)$ be the connection weights from the ith unit on the input layer to the jth unit on the hidden layer, and $w_{jk}^2 (j = 1, 2, ..., J; k = 1, 2, ..., K)$ denote the connection weights from the jth unit on the hidden layer to the kth unit on the output layer. Moreover, $x_i (i = 1, 2, ..., I)$ represent the normalized input values of the ith unit on the input layer, and $y_k (k = 1, 2, ..., K)$ are the output values. Various OSS fault data sets are applied to the input values $x_i (i = 1, 2, ..., I)$.

Considering the number of characteristics for the OSS fault data on BTSs, this chapter applies the following data as parameters to the input data $x_i (i = 1, 2, ..., I)$:

- Date and time registered on BTS
- OSS product name
- OSS component name

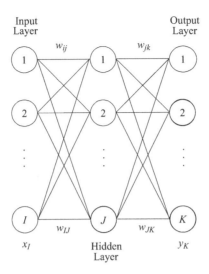

FIGURE 1.1
The framework of our NN by backpropagation learning.

- OSS version name
- Nickname of OSS bug reporter
- Nickname of OSS bug assignee
- OSS bug status
- OS name for each bug

The input and output rules for each unit on each layer are defined by

$$h_j = f\left(\sum_{i=1}^{I} w_{ij}^1 x_i\right),\tag{1.1}$$

$$y_k = f\left(\sum_{j=1}^{J} w_{jk}^2 h_j\right),\tag{1.2}$$

where a logistic function $f(x)$ is a sigmoid function defined as follows:

$$f(x) = \frac{1}{1 + e^{-\theta x}},\tag{1.3}$$

where θ is the gain of sigmoid function. This chapter applies the NNs by backpropagation learning [5] to understand the trend of OSS fault big data. This chapter defines the error function given by

$$E = \frac{1}{2} \sum_{k=1}^{K} (y_k - d_k)^2, \tag{1.4}$$

where $d_k (k = 1, 2, ..., K)$ are the target input values for the output values.

1.4.2 DL Approach

The flow of processing in DL is shown in Figure 1.2. In this figure, $z_l (l = 1, 2, ..., L)$ and $z_m (m = 1, 2, ..., M)$ are the training units in the first layer. Also, $o_n (n = 1, 2, ..., N)$ is the output layer. The output layer refers to the amount of compressed characteristics as the objective variable. Several researchers have proposed the algorithms of DL [6–11]. This chapter focuses on the DL based on feed forward NN to analyze the fault big data on BTSs of OSS [12–17].

This chapter applies the following data sets to the parameters of pretraining units. Then, the objective variable is obtained as the main software component, as shown in Table 1.1. For example, the software managers of cloud computing can obtain two kinds of main software components of the software product from the BTS as the amount of compressed characteristics, that is, OpenStack [18] and Hadoop [19], respectively. This chapter assesses the

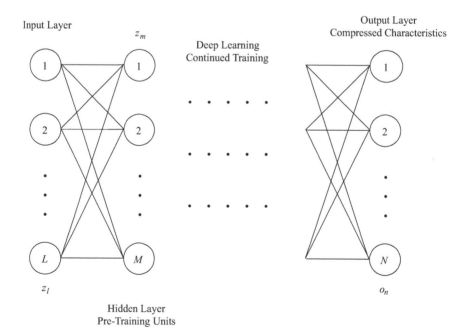

FIGURE 1.2
The framework of our DL.

TABLE 1.1

The Main Software Components for Learning Data in
Case of Three Categories

Index Number	Software Component
1	Other
2	OpenStack
3	Hadoop

effectiveness in case of two main software components considering the practical use of the proposed method [13–15].

- Date and time registered on BTS
- OSS component name
- OSS version number
- Nickname of bug reporter
- Nickname of bug assignee
- OSS bug status
- OS name for each bug
- Bug severity for each bug

1.4.3 Time Series Analysis

The mean time between software failures (MTBF) is well known as the useful measure for assessing the frequency of software failure occurrences. In this chapter, the MTBF for the kth software failure is defined as

$$\lambda_k = \lambda_{k-1} + dl_{k-1} \cdot SD_{k-1},$$ (1.5)

where dl_{k-1} is the $k - 1$ coefficient of reliability trend; for example, $dl_{k-1} = 1$ means the reliability growth and $dl_{k-1} = -1$ the reliability regression. Also, SD_{k-1} is the $k - 1$ standard deviation defined by

$$SD_k = \sqrt{\frac{1}{k-1}\sum_{n=1}^{k}\left(\lambda_n - \frac{1}{n}\sum_{l=1}^{n}\lambda_l\right)^2}.$$ (1.6)

Moreover, the moving average MTBF will be helpful for the OSS developers and managers in assessing the trend of reliability growth. This chapter defines the moving average MTBF as follows:

$$\lambda_l^m = \lambda_{l-1}^m + dl_{l-1}^m \cdot SD_{l-1},$$ (1.7)

where λ_{l-1}^m is the m-fault moving average for $l-1$th actual faults. Also, dl_{l-1}^m is the $l-1$th coefficient of reliability trend estimated by using DL.

Similarly, the maintenance time is well known as the useful measure for understanding the status of software quality. The maintenance time for the kth software failure is given as

$$\mu_k = \mu_{k-1} + dl_{k-1} \cdot SD_{k-1}, \tag{1.8}$$

where dl_{k-1} is the $k-1$ coefficient of maintainability trend; for example, $dl_{k-1} = 1$ means the quality growth and $dl_{k-1} = -1$ the quality regression.

As is the case in MTBF, the moving average maintenance time assists the software project managers in understanding the status of software quality. The moving average maintenance time is defined as follows:

$$\mu_l^m = \mu_{l-1}^m + dl_{l-1}^m \cdot SD_{l-1}, \tag{1.9}$$

where μ_{l-1}^m is the mth moving average maintenance time for the $l-1$th actual number of maintenance. Also, dl_{l-1}^m is the $l-1$th coefficient of maintainability trend estimated by using DL [16,17].

As is the component identification using the DL approach discussed in Section 1.4.2, the framework of the DL in this chapter is shown in Figure 1.2. In Figure 1.2, $z_a(a=1,2,\ldots,A)$ and $z_b(b=1,2,\ldots,B)$ denote the pretraining units. Also, $o_d(d=1,2,\ldots,D)$ is the representation of compressed characteristics. Several DL algorithms have been proposed. This chapter applies the DL based on feed forward NN to learn the fault big data on BTSs of OSS projects. The discussed method of time series analysis applies the following data sets to the parameters of pretraining units. The objective variable is given as the index number dl_i of the ith coefficient to the output values $O_d(d=1,2)$, as shown in Table 1.2. This chapter applies the following data sets as parameters to the input data $x_i(i=1,2,\ldots,I)$:

- Date and time registered on BTS
- Changed date and time of software fault
- Software product
- Software component
- Software version
- Fault reporter for each bug of OSS
- Fault assignee for each bug of OSS
- Status of software bug
- OS for each bug of OSS
- Bug severity level for each bug

TABLE 1.2

The Coefficient Value of Learning Data

Coefficient Value	Reliability Trend
1	Growth
2	Regression

The explanatory variables of ten factors are used for pretraining units. Each data of the explanatory variables is then updated from the character data to the numerical value, such as the occurrence rate.

1.5 Numerical Examples

1.5.1 Component Identification for Software Failure

OpenStack [18] is well known as the cloud software. Also, Hadoop [19] is known as a framework for processing large-scale data sets across clusters of computers. At present, Hadoop is embedded in OpenStack as a component. This chapter focuses on the operating situation by using OpenStack, Hadoop, and other software components. This chapter discusses the use of OpenStack, Hadoop, and the other components

Three kinds of software components on BTSs are considered as the objective variable. The estimation result based on NN of backpropagation learning by using 50% learning data is presented in Figure 1.3. Similarly, the estimation result based on DL by using 50% learning data is presented in

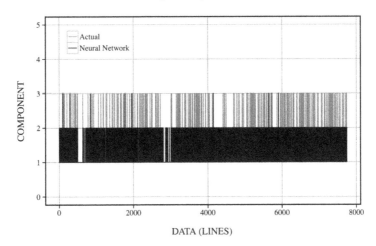

FIGURE 1.3
The estimation result based on NN by using 50% learning data.

Figure 1.4. From Figures 1.3 and 1.4, the estimate based on DL fits better than the one based on NN, for the future in fact. The estimation result based on NN by using 70% learning data is represented in Figure 1.5. Similarly, the estimation result based on DL by using 70% learning data is represented in Figure 1.6. Moreover, the estimation result based on NN and DL by using 90% learning data is represented in Figures 1.7 and 1.8.

Moreover, the comparison results of the recognition rate for NN and DL are shown in Figure 1.9. From this figure, the performance of the estimated recognition rate based on the DL is better than that based on the NN.

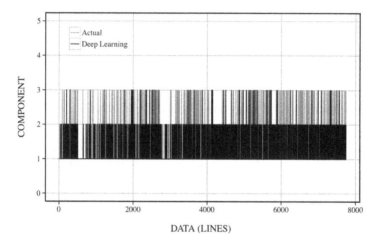

FIGURE 1.4
The estimation result based on DL by using 50% learning data.

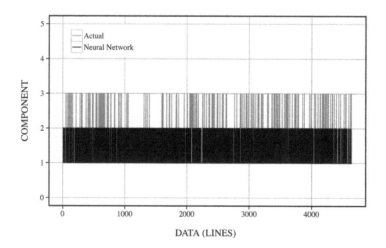

FIGURE 1.5
The estimation result based on NN by using 70% learning data.

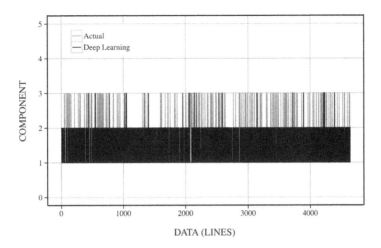

FIGURE 1.6
The estimation result based on DL by using 70% learning data.

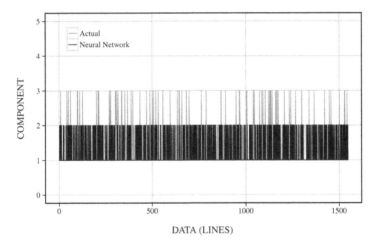

FIGURE 1.7
The estimation result based on NN by using 90% learning data.

1.5.2 Reliability Analysis Based on Time Series

This chapter focuses on an open source HTTP server [20] managed and developed by the Apache HTTP Server Project. The open source Apache HTTP Server runs on modern operating systems such as UNIX and Windows. The Apache HTTP Server Project is one of the most popular software projects of the Apache Software Foundation.

This chapter discussed the practical data sets registered on the BTS of the Apache HTTP Server Project. It analyzed 10,000 fault data sets collected from

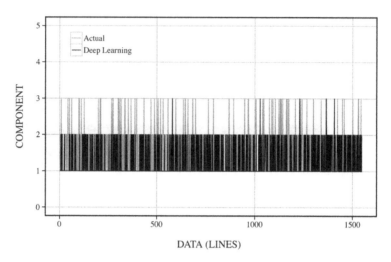

FIGURE 1.8
The estimation result based on DL by using 90% learning data.

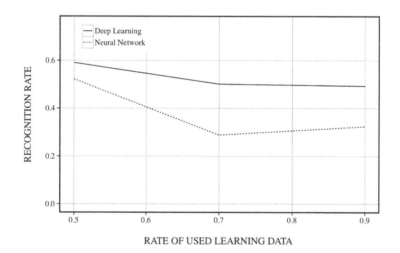

FIGURE 1.9
The comparison results of the recognition rate for NN and DL.

those on the BTS of the Apache HTTP Server Project. It also discussed the opened date and time applied to the objective variable to comprehend the trend of reliability growth from the data registered on the BTS. The estimates of the coefficient value for MTBF based on NN by using 9,000 learning data sets and 1,000 testing data sets are shown in Figure 1.10. Similarly, Figure 1.11 shows the estimates of the coefficient value for MTBF based on DL by using 9,000 learning data sets and 1,000 testing data sets.

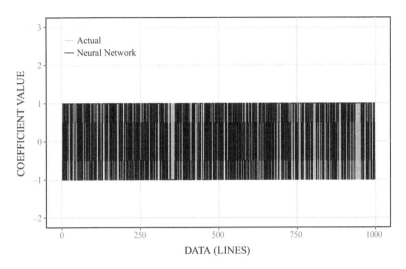

FIGURE 1.10
The estimate of the coefficient value for MTBF based on NN.

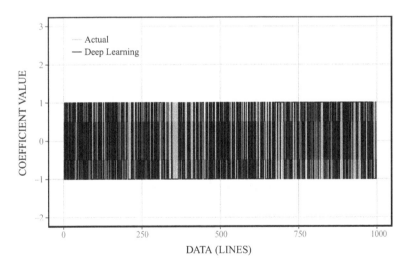

FIGURE 1.11
The estimate of the coefficient value for MTBF based on DL.

Moreover, the estimated MTBF based on DL is illustrated in Figure 1.12, which indicates that the proposed method can accurately estimate the next data line of fault number. Furthermore, Figure 1.13 shows the estimated 50-fault moving average MTBF based on DL. The figure also shows the trend of reliability growth.

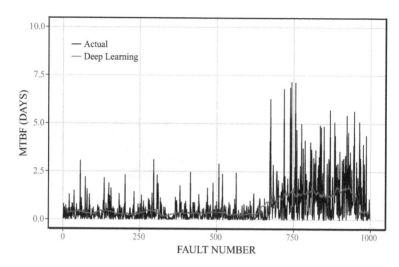

FIGURE 1.12
The estimated MTBF based on DL.

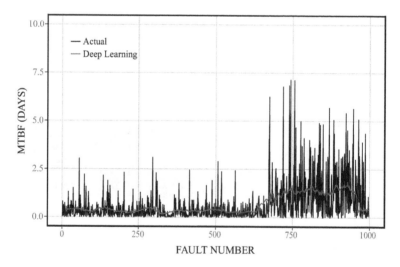

FIGURE 1.13
The estimated 50-fault moving average MTBF based on DL.

1.6 Conclusion

Recently, the BTSs are used in many OSS development and management projects. Also, various fault big data are registered on the BTS. The OSS quality can be improved significantly if the software project managers and developers can make an effective use of fault big data on the BTSs. Using the many data sets on BTSs, the software project managers and developers will be able to take a prompt action for the OSS management and development process, irrespective of whether they can auto-detect for the major component and the major fault.

The OSS fault identification method of the major failure component caused by faults and the method of quality and reliability assessment based on DL by feed forward NN have been discussed in this chapter. In particular, it is difficult to detect the major fault and major component by only the fault count data on BTS. This is because the data contents of bug registered on BTS cannot be guaranteed for the results occurring from the practical OSS, that is, various general users as well as the main development member can report the bug to the BTS. Various numerical illustrations of OSS quality and reliability analysis by using the fault big data in the practical OSS have also been presented in this chapter. Moreover, this chapter has compared the reliability assessment method based on the DL of feed forward with that based on NN by backpropagation learning. Thereby, the method discussed in this chapter can assess OSS quality and reliability in future with high accuracy based on the fault big data on BTS.

In future studies, OSS quality and reliability can be assessed by using various training data sets in practical OSS management and development projects. Therefore, the method based on DL by feed-forward NN is useful to the OSS project developers and managers for assessing the OSS quality and reliability.

Finally, this chapter has introduced the estimation methods of reliability analysis based on AI for OSS. Various conventional models for reliability assessment have been developed by several researchers. However, it is difficult to accurately assess the software quality and reliability by using the fault big data registered on BTS, because of the following reasons:

- Many software reliability models are parametric ones.
- The unknown parameters included in software reliability models are estimated and evaluated by using the fault count data.
- It is very difficult for the OSS project developers and managers to select the optimal model for the practical testing phase of software development projects.

The software reliability can be assessed from various angles, if the software project managers can use all data sets registered on BTS. However, it will be difficult to use all data registered on BTS with such parametric software reliability models. Therefore, this chapter has focused on the DL as AI. This chapter is helpful for the application cases of AI in OSS reliability assessment.

Acknowledgments

This work was supported in part by the JSPS KAKENHI Grant Nos. 15K00102 and 16K01242 in Japan.

References

1. M.R. Lyu, ed., *Handbook of Software Reliability Engineering*, IEEE Computer Society Press, Los Alamitos, CA, 1996.
2. S. Yamada, *Software Reliability Modeling: Fundamentals and Applications*, Springer-Verlag, Tokyo/Heidelberg, 2014.
3. P.K. Kapur, H. Pham, A. Gupta, and P.C. Jha, *Software Reliability Assessment with OR Applications*, Springer-Verlag, London, 2011.
4. S. Yamada and Y. Tamura, *OSS Reliability Measurement and Assessment*, Springer International Publishing, Switzerland, 2016.
5. E.D. Karnin, "A simple procedure for pruning back-propagation trained neural networks," *IEEE Transactions Neural Networks*, Vol. 1, pp. 239–242, 1990.
6. D.P. Kingma, D.J. Rezende, S. Mohamed, and M. Welling, "Semi-supervised learning with deep generative models," *Proceedings of Neural Information Processing Systems*, 2014.
7. B.J. Lafferty, M.R. Rwebangira, and R. Reddy, "Semi-supervised learning using randomized mincuts," *Proceedings of the International Conference on Machine Learning*, Banff, Alberta, Canada, July 4–8, 2004.
8. E.D. George, Y. Dong, D. Li, and A. Alex, "Context-dependent pre-trained deep neural networks for large-vocabulary speech recognition," *IEEE Transactions on Audio, Speech, and Language Processing*, Vol. 20, No. 1, pp. 30–42, 2012.
9. P. Vincent, H. Larochelle, I. Lajoie, Y. Bengio, and P.A. Manzagol, "Stacked denoising autoencoders: Learning useful representations in a deep network with a local denoising criterion," *Journal of Machine Learning Research*, Vol. 11, No. 2, pp. 3371–3408, 2010.
10. H.P. Martinez, Y. Bengio, and G.N. Yannakakis, "Learning deep physiological models of affect," *IEEE Computational Intelligence Magazine*, Vol. 8, No. 2, pp. 20–33, 2013.
11. B. Hutchinson, L. Deng, and D. Yu, "Tensor deep stacking networks," *IEEE Transactions on Pattern Analysis and Machine Intelligence*, Vol. 35, No. 8, pp. 1944–1957, 2013.
12. Y. Tamura and S. Yamada, "Fault identification and reliability assessment tool based on deep learning for fault big data," *Journal of Software Networking*, Vol. 2017, No. 1, pp. 161–176, 2017.
13. Y. Tamura, S. Ashida, M. Matsumoto, and S. Yamada, "Identification method of fault level based on deep learning for open source software," In: R. Lee, ed., *Software Engineering Research, Management and Applications*, Studies in Computational Intelligence, Springer International Publishing, Switzerland, pp. 65–76, 2016.

14. Y. Tamura and S. Yamada, "Comparison of big data analyses for reliable open source software," *Proceedings of the IEEE International Conference on Industrial Engineering and Engineering Management*, Bali, Indonesia, December 4–7, 2016, CD-ROM (Reliability and Maintenance Engineering 3).

15. Y. Tamura, S. Ashida, and S. Yamada, "Fault identification tool based on deep learning for fault big data," *Proceedings of the 3rd International Conference on Information Science and Security*, Pattaya, Thailand, December 19–22, pp. 69–72, 2016.

16. Y. Tamura and S. Yamada, "Reliability analysis based on deep learning for fault big data on bug tracking system," *Proceedings of the International Conference on Reliability, Infocom Technology and Optimization*, Amity University, Noida, India, September 7–9, pp. 37–42, 2016.

17. Y. Tamura and S. Yamada, "Reliability and maintainability analysis and its tool based on deep learning for fault big data," *Proceedings of the International Conference on Reliability, Infocom Technology and Optimization*, Amity University, Noida, India, September 20–22, pp. 104–109, 2017.

18. The OpenStack project, OpenStack, http://www.openstack.org/.

19. The Apache Software Foundation, Apache Hadoop, http://hadoop.apache.org/.

20. The Apache Software Foundation, The Apache HTTP Server Project, http://httpd.apache.org/.

2

Modeling Software Fault Removal and Vulnerability Detection and Related Patch Release Policy

Adarsh Anand and Priyanka Gupta
University of Delhi

Yury Klochkov
St. Petersburg Polytechnic University

V. S. S Yadavalli
University of Pretoria

CONTENTS

2.1 Introduction..19
2.2 Notations...23
2.3 Model Formulation...24
    2.3.1 First Interval: $[0, \tau]$...25
    2.3.2 Second Interval: $[\tau, \tau_p]$...25
    2.3.3 Third Interval: $[\tau_p, T_{LC}]$...26
2.4 Numerical Example..29
2.5 Conclusion 32
References...32

2.1 Introduction

Currently, more and more firms are relying on their IT structure to expand their business, to increase their market share and profitability, and to save money as much as possible. An organization's eagerness to stay ahead of its competitive market leads to the development of advanced software systems. Because of the highly distributed nature of software products, it becomes impractical for the software developers to make such a software framework that entirely satisfies its users' demand. The development of the software is rarely a one-shot game, since it includes risks of dealing with unsatisfied customers by delivering a poor quality product. Therefore, the foremost duty of the software developers is to build full capable software that has a better quality in zero time. Most of the quality issues occur due to defects residing in the software application.

There are many well-known examples of software failures that have caused catastrophic effects. In 2014, Apple's user online data storage was exploited by some hackers that allowed them to post private information of its customers (Kovach, 2014). Due to a software bug, about 100,000 Gmail usernames and their passwords were made public on the Russian forum site, which was later fixed by an update (Kleinman, 2014a). Cyberattacks in the eBay retail company led to the access of contact and log-in details of around 233 million customers, forcing the company to issue a statement requesting its customers to change their details (Nasdaq: EBAY, 2014). Because of a software glitch in Snapchat's third-party application for saving photographs SnapSave, photographs of about 200,000 users were hacked (Kleinman, 2014b). The list of such failures due to this distorted functioning of the software is endless.

The pedagogy that forms the bottom line is all the defects present in the software. Continued testing is what a company can do so as to debug all the defects present in the software. But the query that arises here is how long the testing should be performed. Early researchers have considered the fact that testing of the software should bring to halt with the release of the software. Therefore, for any complex software project, optimal release time determination is one of the most important and crucial decisions to be taken in the testing phase of the software. Project managers have to decide when to stop testing and ship the product to the clients in the market. It can be observed that early release of the software may leave some of the faults and vulnerabilities behind, whereas, with delay in releasing, the company may boost workforce productivity. However, by shipping the software late in the market, management has to bear with contract penalty and loss of market initiative cost.

Literature so far has studied various factors contributing to the testing phase of software and release time of software. Several optimization models have been developed in the field of software reliability. The simplest release time policy was discussed by Goel and Okumoto (1979). They discussed the unconstrained release time problems based on exponential models in two ways—first, with the objective of cost minimization, and second, with the objective of reliability maximization. Yamada and Osaki (1987) discussed the constrained release time problems based on exponential, modified exponential, and s-shaped software reliability growth modeling (SRGM) in two ways—first, with the cost minimization objective under the reliability maximization constraint, and second, with the reliability maximization objective under the cost minimization constraint. Later some researchers also worked on the concepts of penalty cost, testing effort-dependent SRGM, and software random life cycle length, and formulated software release problems based on them (Kapur and Garg, 1989, 1991; Yun and Bai, 1990). Kapur and Garg (1992) incorporated the effect of the removal of dependent faults along with the independent faults in the cost modeling. Pham and Zhang (1999) have further made modifications in the cost modeling by including risk and warranty cost associated with the testing phase of the software life cycle. Researchers have also formulated some of the software release problems incorporating the concepts of imperfect fault debugging, bi-criterion release time problems, error generation, and change point (Kapur and Garg, 1990; Kapur et al., 1994, 2008, 2010). With the evolution of time, new technologies and software were used to find the optimal release time of the software. One such technique was multi-attribute utility theory (MAUT). The first attempt that included the application of MAUT in the release time problem for open source software was made by Xiang li et al. (2011). Singh et al. (2012) introduced a bi-criterion release problem for multi-version software using genetic algorithm.

The threats due to malevolent virus and bugs present in the software are increasing at a fast pace and thus, the software industry professionals are forced to reinvest in their security needs for better development of the company. Sometimes cybercriminals try to exploit vulnerabilities present in the business system to gain access of the corporate IT network. This unauthorized access may lead to the seeping out of sensitive business information, which may result in intense financial losses, thereby ruining the brand reputation. Because of distorted functioning of the software, testing of the software is done extensively for errors that may be introduced during the development.

Even after rigorous testing is done during the different phases of Software Development Life Cycle (SDLC), many defects are exposed only after the large-scale adoption of the software in the real-world scenarios (Arvig, 2016). For combating these errors or attacks and to troubleshoot these problems, a company has to liberate patches for improving their reliability, usability, and system performance. A patch is designed to automate changes in the lines of code involved in implementing of the software. Many theories have stated that if a system wants to be protected against vulnerabilities or faults present in the software, the proficient way to do so is by updating the latest patch in that system (www.ibm.com, 2018). The defects or errors that have not been handled by the extensive amount of testing before product release can be handled in its operations via patches.

The concept of patching helps the firm to increase the duration and efficiency of their testing period. This is because after product release, the users are also working on testing the product. Using the idea of patching, the firms can release their product early and continue to test and improve their product. As soon as any kind of defect is discovered, patches are released for them. This after-sales support leads to satisfied customers and increases the quality of the software.

Various researchers have used their proficiency in fault patching. Anand et al. (2016) have compared the release policy of the software based on two scenarios, that is, with and without patching, incorporating the intensity and efficiency of testers and users. Using the concept of convolution, Das et al. (2015) have developed an optimal cost modeling using the joint effect of users and testers, considering the detection rate of users as constant and testers following exponential distribution. On similar lines, Deepika et al. (2016) developed the mean value function using the joint effect of users and testers considering the detection rate of users and testers following exponential distribution with different rates. Similarly a lot of work has been done in the area of vulnerability detection modeling. On the basis of the learning phenomenon, Alhazmi and Malaiya (2005) have presented revolutionary attempt in the history of vulnerability detection modeling. They developed an S-shaped logistic model to capture vulnerabilities. A multi-release vulnerability model is proposed by Anand et al. (2017), thereby catering to all the leftover vulnerabilities from the preceding software releases. Another study proposed by Anand and Bhatt (2016) captured the hump shape followed by security bugs present in the software. Bhatt et al. (2017) formulated the vulnerability discovery modeling in which additional vulnerabilities were discovered during the life cycle of the software. With the help of a case study, Telang and Wattal (2005) have proposed that if the disclosure of presence of vulnerabilities is made by the management, the stock prices of software vendors may lower down, especially if the patch is not released at the same time. Okamura et al. (2009) discussed the concept of stochastic modeling in determining the optimal patch release without considering vendors and developers' perspective in complete scenario.

Although literature suggests countless optimization models for the optimal release of software and patches, less emphasis is placed on the concept of fault and vulnerability as a single unit. After considering faults and vulnerabilities together in a single patch, the next major decision needed to be taken by the management is when to release the patch into the market. The main concern of the management is to deal with the urgency of patch deployment and schedule its delivery.

After the in-house testing phase of the software, it is released in the market. Until the testing stop time of the software, the presence of any faults or vulnerabilities in the software are mutually detected by testers and users. If their presence is identified, remedial efforts are done by the testers. Further, it is the management's decision to either release the patch for those recovered defects or utilize and save their efforts for the next upgraded version of the software. If the patch is released, detection phenomenon takes place once again and the cycle continues (refer Figure 2.1). In this chapter, a model is proposed on similar lines. A cost modeling is proposed for evaluating software and patch release time using the joint effort of testers and users to detect the defects lying in the software. It is assumed that the fault and vulnerability detection rate of both testers and users are different.

Carrying these assumptions further, the rest of the article is designed as follows: Section 2.2 consists of notations that have been used in the modeling. The proposed modeling framework is presented in Section 2.3. In relation to our proposed modeling, numerical illustration and conclusions are presented in Sections 2.4 and 2.5, respectively.

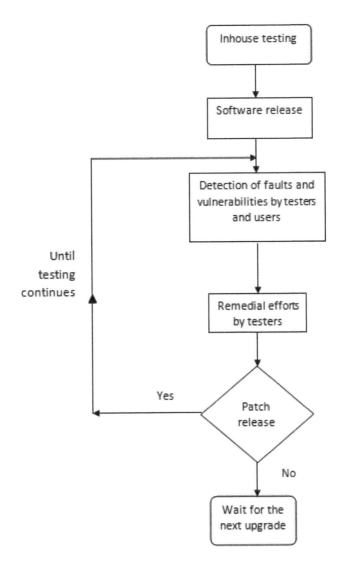

FIGURE 2.1
Flowchart representing life cycle of the software because of patch service.

2.2 Notations

In this section, we have described about the set of notations used for model building.

$m(t)$ Number of faults removed by the time t

$\Omega(t)$ Number of vulnerabilities removed by the time t

a Total number of defects present in the software

τ Release time of the software

τ_p Patch release time

T_{LC} Total life cycle of the software

$C_{x,y}$ Total cost incurred during the time interval $[x, y]$

$\otimes$ Stieltjes convolution

2.3 Model Formulation

In recent years, wide varieties of mathematical models have been developed to predict the reliability and to understand the behavior, structure, and functioning of the software. Among them, NHPP models have proven to be fairly successful. Following their basic assumption, that is, the software is prone to failure due to faults and vulnerabilities present in it, the mean value function representing the SRGM is expressed as

$$\frac{dm(t)}{dt} = \frac{H'(t)}{1 - H(t)}\left(ap_i - m(t)\right) \quad \text{for } i = 1, 2 \tag{2.1}$$

where $H'(t)/1 - H(t)$ represents the defect detection rate or the hazard rate function, and p_1 is the proportion of discovered faults, whereas p_2 is the proportion of discovered vulnerabilities.

By solving Equation (2.1) by means of the initial condition $m(0)=0$, we obtain

$$m(t) = ap_iH(t) \quad \text{for } i = 1, 2 \tag{2.2}$$

This is called the unification scheme (Kapur et al., 2011). It may also be noted that the proportion of faults and vulnerabilities must follow $p_1 + p_2 = 1$. Here different distribution functions will lead to different Mean Value Functions (MVFs).

To build an efficient software system, management has to be very particular about the release time of the software and patches. It has to be chosen in such a way so as to minimize all the faults and vulnerabilities present in the software in the form of security breaches.

According to our proposed model, τ and τ_p are considered the release times of the software and patch, respectively. Therefore, the life cycle of the software $[0, T_{LC}]$ is fragmented into three parts, that is, before release of the software $[0, \tau]$, after release of the software and before patch release $[\tau, \tau_p]$, and the remaining life cycle $[\tau_p, T_{LC}]$ (refer to Figure 2.2).

FIGURE 2.2
Timeline for software and patch releases.

For the proposed modeling framework, the fault removal process is described by the SRGM as given by Goel and Okumoto (1979).

2.3.1 First Interval: $[0, \tau]$

During the in-house testing phase of the software $[0, \tau]$, testers work independently in the system. The testing team has to identify and deal with all the faults present in the software.

Therefore, the total number of faults removed in $[0, \tau]$ is

$$m(\tau) = aH_1(\tau) \tag{2.3}$$

where $H_1(\tau)$ is the fault detection rate, following an exponential distribution. In this phase, we have not contrasted between fault and vulnerability, and therefore, they are removed by the same rate b_1. Therefore,

$$m(\tau) = a\left(1 - e^{(-b_1\tau)}\right) \tag{2.4}$$

Total cost incurred in the fault removal process in this phase is represented by the following expression:

$$C_{0,\tau} = c_1 a\left(1 - e^{(-b_1\tau)}\right) \tag{2.5}$$

where c_1 is the cost associated with defect removal in this interval.

2.3.2 Second Interval: $[\tau, \tau_p]$

In this field testing phase of the software, technical discussion is required to understand the defect removal process. In the proposed modeling, it is assumed that the software is released at the time point τ, but the testing continues throughout the life cycle of the software. The cost of discovering vulnerability is supposed to be higher than the cost of discovering faults. Therefore, in this interval, a distinction between faults and vulnerabilities is considered.

After the release of the software, both testers and users work mutually to detect faults and vulnerabilities present in the software. As soon as a fault or vulnerability is reported by the user, it is removed by the tester. To study this conjoint effort of testers and users in the defect detection phenomenon, the Stieltjes convolution probability function is used (Deepika et al., 2016). The rates of fault and vulnerability detection are considered to be different for testers and users.

The total number of faults removed in the interval $[\tau, \tau_p]$ is represented as

$$m(\tau_p - \tau) = p_1 a\left(1 - H_1(\tau)\right)\left(H_2 \otimes U_2(\tau_p - \tau)\right) \tag{2.6}$$

The total number of vulnerabilities removed in this interval is represented as

$$\Omega(\tau_p - \tau) = (1 - p_1)a\left(1 - H_1(\tau)\right)\left(H_3 \otimes U_3(\tau_p - \tau)\right) \tag{2.7}$$

where $H_2 \otimes U_2(\tau_p - \tau)$ and $H_3 \otimes U_3(\tau_p - \tau)$ are the conjoint fault and vulnerability removal rates for this interval, respectively. H_2 and U_2 are the fault detection rates of testers and users, respectively; H_3 and U_3 are the vulnerability detection rates of testers and users, respectively. The residual number of faults and vulnerabilities of the preceding interval $[0, \tau]$ that are reported in this interval are represented by $p_1 a(1 - H_1(\tau))$ and $(1 - p_1)a(1 - H_1(\tau))$, which are removed with the new fault detection and vulnerability detection rates, that is, with $H_2 \otimes U_2(\tau_p - \tau)$ and $H_3 \otimes U_3(\tau_p - \tau)$, respectively. We have considered that H_2 and H_3 follow exponential distribution with parameters b_2 and b_3, respectively, and U_2 and U_3 are constants.

Therefore, the total number of faults removed is expressed as

$$m(\tau_p - \tau) = p_1 a e^{-(b_1 \tau)}\left(1 - e^{\left(-b_2(\tau_p - \tau)\right)}\right) \tag{2.8}$$

Therefore, the total number of vulnerabilities removed is expressed as

$$\Omega(\tau_p - \tau) = (1 - p_1)a(e^{-(b_1 \tau)})\left(1 - e^{\left(-b_3(\tau_p - \tau)\right)}\right) \tag{2.9}$$

The total cost incurred in this interval to remove all detected faults and vulnerabilities is represented as

$$C_{\tau, \tau_p} = c_2 p_1 a e^{-(b_1 \tau)}\left(1 - e^{\left(-b_2(\tau_p - \tau)\right)}\right) + c_3(1 - p_1)a(e^{-(b_1 \tau)})\left(1 - e^{\left(-b_3(\tau_p - \tau)\right)}\right) \tag{2.10}$$

where c_2 and c_3 are the costs associated with fault and vulnerability removal, respectively.

2.3.3　Third Interval: $[\tau_p, T_{LC}]$

This time period is also known as software's saturation period for the reason that testing of the software continues till its total cycle length, but after the time point τ_p, no more patches are released into the market. Here it is considered that a single patch has the proficiency to fix both faults and vulnerabilities together, which are released in the market at the time τ_p.

The total number of faults removed in this interval is represented as

$$m(T_{LC} - \tau_p) = p_1 a \left(1 - H_1(\tau)\right)\left(1 - (H_2 \otimes U_2)(\tau_p - \tau)\right)\left(H_4 \otimes U_4(T_{LC} - \tau_p)\right) \quad (2.11)$$

The total number of vulnerabilities removed in this interval is represented as

$$\Omega(T_{LC} - \tau_p) = (1 - p_1)a \left(1 - H_1(\tau)\right)\left(1 - (H_3 \otimes U_3)(\tau_p - \tau)\right)\left(H_5 \otimes U_5(T_{LC} - \tau_p)\right)$$

$$(2.12)$$

where $H_4 \otimes U_4(T_{LC} - \tau_p)$ and $H_5 \otimes U_5(T_{LC} - \tau_p)$ are the conjoint fault and vulnerability removal rates for this interval, respectively. H_4 and U_4 are the fault detection rates of testers and users, respectively, and H_5 and U_5 are the vulnerability detection rates of testers and users, respectively. The residual number of faults and vulnerabilities of the interval $[0, \tau_p]$ that are reported in this interval is represented by $p_1 a \left(1 - H_1(\tau)\right)\left(1 - (H_2 \otimes U_2)(\tau_p - \tau)\right)$ and $(1 - p_1)a \left(1 - H_1(\tau)\right)\left(1 - (H_3 \otimes U_3)(\tau_p - \tau)\right)$, respectively, which are to be dealt with the new fault detection and vulnerability detection rates, that is, with $H_4 \otimes U_4(T_{LC} - \tau_p)$ and $H_5 \otimes U_5(T_{LC} - \tau_p)$, respectively.

We have considered H_4 and H_5 follows exponential distribution with parameter b_4 and b_5 and U_4 and U_5 as constant.

Therefore, the total number of faults removed in this interval is expressed as

$$m(T_{LC} - \tau_p) = p_1 a e^{-(b_1 \tau)} e^{\left(-b_2 (\tau_p - \tau)\right)} \left(1 - e^{\left(-b_4 (T_{LC} - \tau_p)\right)}\right) \quad (2.13)$$

Therefore, the total number of vulnerabilities removed is expressed as

$$\Omega(T_{LC} - \tau_p) = (1 - p_1)a e^{-(b_1 \tau)} e^{\left(-b_3 (\tau_p - \tau)\right)} \left(1 - e^{\left(-b_5 (T_{LC} - \tau_p)\right)}\right) \quad (2.14)$$

Hence, the total cost incurred in this case is represented as

$$C_{\tau_p, T_{LC}} = c_4 . p_1 . a . e^{-(b_1 . \tau)} . e^{(-b_2 . (\tau_p - \tau))} . \left(1 - e^{(-b_4 . (T_{LC} - \tau_p))}\right) +$$

$$c_5 (1 - p_1) . a . e^{-(b_1 . \tau)} . e^{(-b_3 . (\tau_p - \tau))} . \left(1 - e^{(-b_5 . (T_{LC} - \tau_p))}\right)$$

$$(2.15)$$

where c_4 and c_5 are the costs associated with fault and vulnerability removal, respectively.

According to our assumption that the testing of the software is continued throughout its total life cycle length, the cost incurred in this process is represented as

$$C_{0, T_{LC}} = c_6 T_{LC} \quad (2.16)$$

where c_6 represents the testing per unit cost.

Combining Equations (2.5), (2.10), (2.15), and (2.16), the total cost incurred in the removal of faults and vulnerabilities together throughout the life time of the software is represented as

$$TC = c_1 a\left(1 - e^{(-b_1 \tau)}\right) + c_2 p_1 a e^{-(b_1 \tau)}\left(1 - e^{\left(-b_2(\tau_p - \tau)\right)}\right) + c_3(1 - p_1)a\left(e^{-(b_1 \tau)}\right)\left(1 - e^{\left(-b_3(\tau_p - \tau)\right)}\right)$$

$$+ c_4 p_1 a e^{-(b_1 \tau)} e^{\left(-b_2(\tau_p - \tau)\right)}\left(1 - e^{\left(-b_4(T_{LC} - \tau_p)\right)}\right) + c_5(1 - p_1)a e^{-(b_1 \tau)} e^{\left(-b_3(\tau_p - \tau)\right)}$$

$$\left(1 - e^{\left(-b_5(T_{LC} - \tau_p)\right)}\right) + c_6 T_{LC}$$

$$(2.17)$$

The management ensures that the total cost incurred in the defect removal phenomenon (represented in Equation (2.17)) is minimized, thereby taking care of the budgetary constraint of the company, the management must continue testing the software for the interim duration of time. Doing so, the management is likely to increase the risk of presence of defects in the software, resulting in the decline in the value of its reliability, which cannot be overlooked. Therefore, the release time of the software and patch must be chosen in such a way so as to strike the balance between this incurred cost and level of the reliability.

Therefore, an optimization problem (P1) is developed considering the budgetary and reliability constraint, which is represented as follows:

$$Min.TC(\tau, \tau_p) = c_1 a\left(1 - e^{(-b_1 \tau)}\right) + c_2 p_1 a e^{-(b_1 \tau)}\left(1 - e^{\left(-b_2(\tau_p - \tau)\right)}\right)$$

$$+ c_3(1 - p_1)a\left(e^{-(b_1 \tau)}\right)\left(1 - e^{\left(-b_3(\tau_p - \tau)\right)}\right) + c_4 p_1 a e^{-(b_1 \tau)} e^{\left(-b_2(\tau_p - \tau)\right)}$$

$$\times\left(1 - e^{\left(-b_4(T_{LC} - \tau_p)\right)}\right) + c_5(1 - p_1)a e^{-(b_1 \tau)} e^{\left(-b_3(\tau_p - \tau)\right)} \qquad (2.18)$$

$$\times\left(1 - e^{\left(-b_5(T_{LC} - \tau_p)\right)}\right) + c_6 T_{LC}$$

$$s.t.\ TC(\tau, \tau_p) \le C_b \quad R \ge r$$

where C_b is the maximum budget allocated by the management for the project, R is the acquired reliability by the software, and r is the minimum reliability level required.

Mathematically, this risk or the attribute of reliability can be represented as the ratio of total number of faults and vulnerabilities removed from the software to the total number of defects present in it in the beginning.

$$R = \frac{m(T_{LC})}{a}$$

where, $m(T_{LC})$ is the total number of faults and vulnerabilities removed from the software.

Total number of faults and vulnerabilities removed from the software in the particular interval is represented in Table 2.1. Therefore, summing up the final defect removal content, $m(T_{LC})$ can be represented as

TABLE 2.1

Total Number of Faults and Vulnerabilities Removed

Interval	Total Number of Faults	Total Number of Vulnerabilities
$[0, \tau]$	$a\left(1 - e^{(-b_1 \tau)}\right)$	–
$[\tau, \tau_p]$	$p_1 a e^{-(b_1 \tau)}\left(1 - e^{\left(-b_2(\tau_p - \tau)\right)}\right)$	$(1 - p_1) a \left(e^{-(b_1 \tau)}\right)\left(1 - e^{\left(-b_3(\tau_p - \tau)\right)}\right)$
$[\tau_p, T_{LC}]$	$p_1 a e^{-(b_1 \tau)} e^{\left(-b_2(\tau_p - \tau)\right)}\left(1 - e^{\left(-b_4(T_{LC} - \tau_p)\right)}\right)$	$(1 - p_1) a e^{-(b_1 \tau)} e^{\left(-b_3(\tau_p - \tau)\right)}\left(1 - e^{\left(-b_5(T_{LC} - \tau_p)\right)}\right)$

$$m(T_{LC}) = a\left(1 - e^{(-b_1 \tau)}\right) + p_1 a e^{-(b_1 \tau)}\left(1 - e^{\left(-b_2(\tau_p - \tau)\right)}\right) + (1 - p_1) a\left(e^{-(b_1 \tau)}\right)\left(1 - e^{\left(-b_3(\tau_p - \tau)\right)}\right)$$

$$+ p_1 a e^{-(b_1 \tau)} e^{\left(-b_2(\tau_p - \tau)\right)}\left(1 - e^{\left(-b_4(T_{LC} - \tau_p)\right)}\right) + (1 - p_1) a e^{-(b_1 \tau)} e^{\left(-b_3(\tau_p - \tau)\right)}$$

$$\left(1 - e^{\left(-b_5(T_{LC} - \tau_p)\right)}\right)$$

The formulated optimization model (P1) can be solved using mathematical tools (Lingo, Mathematica) to determine the value of software and patch release time.

2.4 Numerical Example

For the validation of the proposed methodology, an empirical numerical analysis is conducted. We have considered the real-life data set of 1,500 telephone subscribers of TROPICO R-1500 switching systems (Kanoun et al., 1991). A total of 461 faults were removed during its execution period of 81 weeks (refer Table 2.2). Further, its execution period of 81 weeks is divided into three phases—a time period of 1–30 weeks represents the validation phase, a time period of 31–42 weeks represents the field trial phase, and the remaining time period of 43–81 weeks represents the system operation phase.

With the help of the method of nonlinear Ordinary Least Square (OLS), estimation for the parameters of the proposed SRGM is done by using Statistical Analysis System (SAS). The parameter estimation values are presented in Table 2.3. After substituting the value of parameters, the number of faults and vulnerabilities in any interval (modeling given in Table 2.1) is represented in Table 2.4. After examining the estimated value of parameter p_1, it is observed that faults constitute the major proportion (89.9%) of defects lying in the software. A slight decrease in the efficiency of testers to identify the existence of faults before and after release of the software (b_1 and

TABLE 2.2

Failure Count Data

Time Frame	Fault Count	Time Frame	Fault Count	Time Frame	Fault Count	Time Frame	Fault Count
1	7	22	214	43	356	64	423
2	8	23	223	44	367	65	429
3	36	24	246	45	373	66	440
4	45	25	257	46	373	67	443
5	60	26	277	47	378	68	448
6	74	27	283	48	381	69	454
7	80	28	286	49	383	70	456
8	98	29	292	50	384	71	456
9	106	30	297	51	384	72	456
10	115	31	301	52	387	73	457
11	120	32	302	53	387	74	458
12	134	33	310	54	387	75	459
13	139	34	317	55	388	76	459
14	142	35	319	56	393	77	459
15	145	36	323	57	398	78	460
16	153	37	324	58	400	79	460
17	167	38	338	59	407	80	460
18	174	39	342	60	413	81	461
19	183	40	345	61	414		
20	196	41	350	62	419		
21	200	42	352	63	420		

TABLE 2.3

Parameter Estimates

Parameters	Values
A	681.486
b_1	0.018
b_2	0.0103
b_3	0.0958
b_4	0.0103
b_5	0.066
p_1	0.899

b_2, respectively) is also observed. Therefore, it validates the fact that before putting the software into operational use, rigorous testing of the software is done to remove maximum number of faults from the software. This can be verified from Table 2.4.

It can also be interpreted that as soon as the software is in the market, rigorous testing should be done to remove vulnerabilities present in the system,

TABLE 2.4

Total Defect Content in the Respective Interval

Interval	Number of Removed Faults	Number of Removed Vulnerabilities
$[0, \tau]$	294.2409	–
$[\tau, \tau_p]$	59.286	32.221
$[\tau_p, T_{LC}]$	117.1014	6.643

thereby making an attempt to free the software from any security breaches as soon as possible. With the course of time, slender decrease in the vulnerability removal rate is observed (b_3 and b_5, respectively). This can also be verified from Table 2.4, which demonstrates that the number of vulnerabilities removed before the patch is in the market is higher compared to that after its release.

The comparison criteria for goodness of fit for the data set for the proposed model are represented in Table 2.5. The quality of our proposition is assessed by mean square error (MSE) and root MSE (RMSE), whose lower values claim the better fit. Moreover, higher values of R^2 and Adj R^2 point toward the accuracy of the modeling.

Further, to obtain the release time of the software and the patch, values of some of the parameters are assumed as constant. The total life cycle of the software is assumed to be of 100 weeks and the maximum budget allocated by the management for the development of the software is $20,000. The minimum reliability level requirement of the software is assumed to be 0.6. The hypothetical values of the cost parameters are assumed to be $c_1 = \$20$, $c_2 = \$26$, $c_3 = \$29$, $c_4 = \$23$, $c_5 = \$43$, and $c_6 = \$15$.

Using the software package Lingo, problem (P1) has been solved to minimize the total cost incurred during the defect removal process under budgetary and reliability constraints. The optimal release time of the software and the patch are 31.401 weeks and 49.526 weeks, respectively. The optimal value of the objective function is $12,844.87.

From the results obtained after solving the proposed modeling framework, it can be concluded that the time to release the software coincides with the time point at which the software is offered for its field trials. After completing the field trail, the software continues its operations and a certain number of users continue to use the software. To remove any faults and vulnerabilities

TABLE 2.5

Goodness of Fit

Comparison Criteria	Values
MSE	74.5279
RMSE	8.633
R^2	0.9959
Adj R^2	0.9958

that were encountered during the field trial (after the software is released), the software vendor has to offer a patch on 49 weeks to minimize any risk involved in a breach.

Hence the outcome implies that the proposed methodology captures the real-life scenario in a more relevant manner. On the basis of our results, we can conclude that it is advisable for software development firms to engage with post-release testing activities.

2.5 Conclusion

According to the well-known fact, software development is a time- and effort-consuming procedure. The pedagogy that forms the bottom line here is all the remaining faults and vulnerabilities present in the software even after the testing phase of the software. Due to this, software producers can offer updates using patches. They can choose to release the software early in the market rather than at the time recommended by the traditional approach and continue to test the software for longer duration of time.

Software reliability and its quality are dependent on the testing team of the software. Testers debug the software before its release. After its release, testers and users jointly detect the faults and vulnerabilities present in the software that are therefore removed by the testing team. Hence, a combined patch dealing with the faults and vulnerabilities present in the software is released in the market. On the basis of the aforesaid modified approach, in this study, optimal release time and patch release time of the software have been determined. A nonlinear optimization problem is developed for the same under the budgetary and reliability constraints.

For future context, rather than using exponential distribution, we also intend to study the use of more generic distributions. Different forms of distribution functions $F(t)$ and $U(t)$ will alter the value of mean value function. It will be interesting to observe the changes in their results.

References

Alhazmi O.H., Malaiya Y.K. (2005), Modeling the vulnerability discovery process, In *ISSRE 2005: 16th IEEE International Symposium on Software Reliability Engineering*, pp. 10.

Anand. A, Bhatt N. (2016), Vulnerability discovery modeling and weighted criteria based ranking, *J Indian Soc Prob Stat*, 17(1):1–10.

Anand A., Agarwal M., Tamura Y., Yamada S. (2016), Economic impact of software patching and optimal release scheduling, *Qual Reliab Engng Int.* 33(1): 149–157.

Anand A., Das S., Aggrawal D., Klochkov Y. (2017), Vulnerability discovery modeling for software with multi-versions, In Mangey Ram and J. Paulo Davim (Eds) *Advances in Reliability and System Engineering*, Springer International Publishing, Switzerland, pp. 255–265.

Arvig (2016), The value of software patches and updates for your business, www.arvigbusiness.com/the-value-of-software-patches-and-updates-for-your-business/, Accessed date: 29-Jan-2018.

Bhatt N., Anand A., Yadavalli V.S.S., Kumar V. (2017), Modeling and characterizing software vulnerabilities, *Int J Math Eng Manag Sci*, 2(4):288–299.

Das S., Anand A., Singh O., Singh J. (2015), Influence of patching on optimal planning for software release and testing time, *CDQM*, 18(4):81–92.

Deepika, A.A., Singh N., Dutt P. (2016), Software reliability modeling based on in-house and field testing, *CDQM*, 19(1), 74–84.

Goel A.L., Okumoto K. (1979), Time-dependent error detection rate model for software reliability and other performance measures, *IEEE Trans Reliab*, 28(3):206–211.

IBM Knowledge Center (2018), Why patch management is important, www.ibm.com/support/knowledgecenter/en/SSTFWG_4.3.1/com.ibm.tivoli.itcm.doc/CMPMmst20.htm, Accessed date: 29-Jan-2018.

Kanoun K., Martini M., Souja J. (1991), A method for software reliability analysis and prediction application to the TROPICO-R switching systems, *IEEE Trans Softw Eng*, 17(4):334–344.

Kapur P.K., Garg R.B. (1989), Cost-reliability optimal release policies for software system under penalty cost, *Int J Syst Sci*, 20:2547–2562.

Kapur P.K., Garg R.B. (1990), Optimal software release policies for software reliability growth models under imperfect debugging, *Recherchè Operationanelle/Oper Res*, 24:295–305.

Kapur P.K., Garg R.B. (1991), Optimal release policies for software systems with testing effort, *Int J Syst Sci*, 22(9):1563–1571.

Kapur P.K., Garg R.B. (1992), A software reliability growth model for an error removal phenomenon, *Softw Eng J*, 7:291–294.

Kapur P.K., Aggarwal S., Garg R.B. (1994), Bi-criterion release policy for exponential software reliability growth models, *Recherchè Operationanelle/Oper Res*, 28:165–180.

Kapur P.K., Gupta D., Gupta A., Jha P.C. (2008), Effect of introduction of fault and imperfect debugging on release time, *J Ratio Math*, 18:62–90.

Kapur P.K., Garg R.B., Aggarwal A.G., Tandon A. (2010), General framework for change point problem in software reliability and related release time problem, In *Proceedings ICQRIT 2009*.

Kapur P.K., Pham H., Gupta A., Jha P.C. (2011), *Software Reliability Assessment with OR Application*, Springer, Berlin.

Kleinman A. (2014a), 5 million gmail usernames and associated passwords leaked, *The Huffington Post*, September 11, 2014, www.huffingtonpost.com/2014/09/11/gmail-passwords-hacked_n_5805104.html, Accessed date: 30-Jan-2018.

Kleinman A. (2014b), 200,000 Snapchat photos leaked on 4Chan, *The Huffington Post*, October 10, 2014, www.huffingtonpost.com/2014/10/10/snapchat-leak_n_5965590.html, Accessed date: 30-Jan-2018.

Kovach S. (2014), We still don't have assurance from Apple that iCloud is safe, *Business Insider*, September 2, 2014, www.businessinsider.com/apple-statement-on-icloud-hack-2014-9, Accessed date: 30-Jan-2018.

Li X., Li Y.F., Xie M., Ng S.H. (2011), Reliability analysis and optimal version-updating for open source software. *Information and Software Technology*, 53(9): 929–936.

Nasdaq: EBAY (2014), News release, eBay Inc. to Ask eBay Users to Change Passwords, eBay, May 21, 2014, www.ebayinc.com/in_the_news/story/ebay-inc-ask-ebay-users-change-passwords, Accessed date: 30-Jan-2018.

Okamura H., Tokuzane M., Dohi T. (2009), Optimal security patch release timing under non-homogeneous vulnerability-discovery process. In *2009 20th International Symposium on Software Reliability Engineering*, November, pp. 120–128, IEEE.

Pham H., Zhang X. (1999), A software cost model with warranty and risk costs. *IEEE Transactions on Computers*, 48(1): 71–75.

Singh O., Kapur P.K., Anand A. (2012), A multi attribute approach for release time and reliability trend analysis of a software, *Int J Syst Assurance Eng Manage*, 3(3):246–254.

Telang R., Wattal S. (2005), Impact of vulnerability disclosure on market value of software vendors. An empirical analysis, In *Proceedings of the Fourth Annual Workshop on Economics and Information Security (WEIS06)*.

Yamada S., Osaki S. (1987), Optimal software release policies with simultaneous cost and reliability requirement, *Eur J Oper Res*, 31:46–51.

Yun W.Y., Bai D.S. (1990), Optimal software release policies with random life cycle, *IEEE Trans Reliab*, 39(2):167–170.

3

System Reliability Optimization in a Fuzzy Environment via Hybridized GA–PSO

Laxminarayan Sahoo

Raniganj Girls' College

CONTENTS

3.1   Introduction..35
3.2   Some Mathematical Notions..37
    3.2.1   Introductory Definitions about Fuzzy Set....................................37
    3.2.2   Beta Distribution...38
    3.2.3   Expected Value/Mean Value of Beta Distribution.....................38
    3.2.4   Defuzzification of TFN Constructed from Statistical Beta
        Distribution..38
3.3   Notations...39
3.4   Mathematical Formulation of RRAP...39
3.5   Constraint-Handling Technique for Constrained Mixed-Integer
    Nonlinear Problems 40
3.6   Solution Methodology 41
    3.6.1   Genetic Algorithm 42
    3.6.2   Particle Swarm Optimization 43
        3.6.2.1   Constriction Coefficient–Based PSO 44
3.7   Numerical Examples 44
3.8   Concluding Remarks..47
References...47

3.1 Introduction

Development of a technological system, the decorative pattern of the system, and system reliability are indispensable measures in industries, especially in complicated manufacturing systems, namely, smartphone, laptop, digital laboratory instruments, and smart watch. The purpose of the reliability optimization problem is to upgrade the reliability of a system within given budget constraints. Reliability redundancy allocation problem (RRAP) aims at finding the number of redundant components and also the reliability of each component to maximize the system reliability and/or minimize the system cost/system volume/system weight under requisite budget constraints. RRAP, mostly a nonlinear mixed-integer programming problem, cannot be solved easily because of the discrete domain of the problem. According to Chern [1], RRAP is an NP-hard problem. Regarding reliability optimization, there are basically four types of system configurations available in the literature, namely, series, parallel, series parallel, and complicated. In RRAP it is seen that an unspecified number of decision variables are integer types and the others are continuous types. Therefore, finding optimal solutions to such types of optimization problems is a very hard job for the reliability designers. In the past, several methods, like heuristic methods [2–6], reduced gradient method [7], branch and bound method [8–11], integer programming [12], dynamic programming [13,14], and distinct well-accepted optimizing models were primarily the tools for solving such redundancy allocation problems. But these approaches, along with their merits, have demerits too. In a few particular configurations of system reliability design, the corresponding reliability functions are not decomposable, and for this reason, the well-known optimization technique, namely, dynamic programming, is not useful in solving such problems. The branch-and-bound programming method is not useful, because in this approach, the efficacy depends on precision of the bound of the variables/parameters and the required memory grows up exponentially with the dimension of the problem. Hence, heuristic approaches, namely, genetic algorithm (GA), particle swarm optimization (PSO), and differential evolution (DE), give a reasonable solution to such complicated optimization problems within a permissible time complexity as well as space complexity, and to upgrade computational efficiency, researchers have used hybridized algorithms to meet their goals. In this connection, GA, ant colony optimization, simulated annealing (SA), PSO, and DE have successfully been considered as optimization tools for solving reliability optimization problems. To hybridize, GA has been assembled with other heuristic algorithms to meet the computational efficiency.

The issues of application of GAs in reliability optimization problems have been dealt with by Dengiz et al. [15], Tavakkoli-Moghaddama et al. [16], Ye et al. [17], Gupta et al. [18], Bhunia et al. [19,20], Sahoo et al. [21,22], Mahato et al. [23], and Sahoo et al. [24,25]. Genetic algorithm has also been observed in the work of Hsieh et al. [26]. They solved mixed-integer nonlinear reliability optimization problems considering system configurations, namely, series, parallel, series–parallel, and complex. At the same time, Kuo et al. [27] studied the application of SA in reliability optimization. Kim et al. [28] solved RRAP using SA algorithm and unveiled that their result is much better in contrast to the results given by Hitika et al. [14], Hsieh et al. [26], and Yokota et al. [29]. Related work has been relayed in the existing literature taking series and complex (bridge) systems configuration using PSO [30].

In the previous studies, the design parameters of RRAP were ordinarily considered to be error-free values. However, in real-life conditions, it is not possible to get the exact values of the parameters, and hence, the design parameters are considered as an ambiguous value. In such situations, the fuzzy set theory is the most well-established tool to manage such ambiguous/imprecise parameters [31]. For reliability optimization with ambiguous parameters, the works of Sahoo et al. [21] and Bhunia et al. [19] are a good source of information.

The purpose of this chapter is to refine an efficient algorithm on the basis of GA and PSO for solving mixed-integer nonlinear reliability optimization problems in complicated reliability systems with ambiguous/imprecise parametric values. Here, we have treated ambiguous parameters considering fuzzy number. Therefore, the fuzzy valued RRAP yields an efficient setup that solves the complicated RRAP with fuzzy parameters. In this chapter, we have taken into consideration the RRAP of complicated systems. In this chapter, the reliability of each component and other parameters related to this problem have been considered to be a triangular fuzzy number (TFN), and reconstruction of equivalent problem has been done to crisp optimization problem with the help of defuzzification method [32]. Here, a new defuzzification method on the basis of beta distribution of first kind [33] has been applied for the defuzzification of fuzzy numbers. Big-M penalty technique [18] has been used to reshape the constructed problem into an unconstrained optimization problem, and the same has been solved by our refined hybridized GA–constriction coefficient-based PSO (PSO-Co) algorithm. Finally, to test the effectiveness of the hybridized algorithm, a numerical example has been considered and solved.

3.2 Some Mathematical Notions

3.2.1 Introductory Definitions about Fuzzy Set

A fuzzy set $\tilde{A}$ of universe X is defined by function $\mu_{\tilde{A}}(x)$, which corresponds to each member x in X to a real number in the interval $[0,1]$, where $\mu_{\tilde{A}}(x)$ is called the membership function of the set $\tilde{A}$.

Definition 3.1 The α-cut set defined as $\tilde{A}_{\alpha} = \{x \in X : \mu_{\tilde{A}}(x) \geq \alpha\}$, where $\alpha \in [0,1]$.

Definition 3.2 A fuzzy set $\tilde{A}$ is called normal if $\mu_{\tilde{A}}(x) = 1$ for at least one point $x \in X$.

Definition 3.3 Let $x_1, x_2 \in X$ and $\lambda \in [0,1]$ if $\mu_{\tilde{A}}(\lambda x_1 + (1-\lambda)x_2) \geq \min \{\mu_{\tilde{A}}(x_1), \mu_{\tilde{A}}(x_2)\}$ then $\tilde{A}$ is called convex fuzzy set.

Definition 3.4 A convex and normal fuzzy set is called fuzzy number.

Definition 3.5 A TFN is denoted as $\tilde{A} = (a_1, a_2, a_3)$ and is defined by

$$\mu_{\tilde{A}}(x) = \begin{cases} \dfrac{x - a_1}{a_2 - a_1} & \text{if } a_1 \le x \le a_2 \\ 1 & \text{if } x = a_2 \\ \dfrac{a_3 - x}{a_3 - a_2} & \text{if } a_2 \le x \le a_3 \end{cases}$$

3.2.2 Beta Distribution

The probability density function of a random variable X that follows a beta distribution is given by

$$f(x) = \begin{cases} \dfrac{x^{p-1}(1-x)^{q-1}}{B(p,q)} & 0 < x < 1 \\ 0 & \text{otherwise} \end{cases}$$

where $p > 0$ and $q > 0$ are the parameters of beta distribution.

3.2.3 Expected Value/Mean Value of Beta Distribution

The expected value or mean of beta distribution is $p/(p+q)$. If the density curve of beta distribution is unimodal and let $x^* \in (0,1)$ be the point at which the density curve $f(x)$ gave its maximum value, then $q - 1 = (p-1)\left(1 - x^*/x^*\right)$.

So, for both the values of p and x^*, the value of q can be calculated, and the mean of beta distribution can be obtained by $\mu = p/(p+q)$.

3.2.4 Defuzzification of TFN Constructed from Statistical Beta Distribution

If $\tilde{A} = (a_1, a_2, a_3)$ is a TFN, then the projection of $\tilde{A} = (a_1, a_2, a_3)$ on the interval $(0,1)$ is $\tilde{A} = \left(0, (a_2 - a_1)/(a_3 - a_1), 1\right)$. Now, if we take $p = (a_2 - a_1)/(a_3 - a_1) + 1$, then $q = (a_3 - a_2)/(a_3 - a_1) + 1$, and hence, $\mu = (a_3 + a_2 - 2a_1)/3(a_3 - a_1)$. Now, $\mu_{\tilde{A}}$ is a real-valued number; by altering μ from the interval $(0,1)$ to the interval (a_1, a_3), it is deemed a real-valued number, which corresponds to the following fuzzy number $\tilde{A} = (a_1, a_2, a_3)$:

$$\mu_{\tilde{A}} = \mu(a_3 - a_1) + a_1 = \frac{a_3 + a_2 + a_1}{3}$$

Remark 3.1 Using beta distribution, the crisp real-valued number $\mu_{\tilde{A}}$ corresponds to the TFN $\tilde{A} = (a_1, a_2, a_3)$ is $\mu_{\tilde{A}} = (a_1 + a_2 + a_3)/3$.

3.3 Notations

Some notations that are used in the entire chapter are presented in Table 3.1.

3.4 Mathematical Formulation of RRAP

The mathematical formulation of RRAP is given as follows:

$$\text{Maximize } R_S = f(x, r) \tag{3.1}$$

subject to $g_j(x, r) \le b_j, \quad j = 1, 2, \ldots, m$

$(x, r) = (x_1, x_2, \ldots, x_n, r_1, r_2, \ldots, r_n), \quad 1 \le l_i \le x_i \le u_i$ and $x_i \in \mathbb{Z}^+, \quad 0 < \tilde{r}_i < 1$ and $i = 1, 2, \ldots, n$.

where R_S is the system reliability, $g_j(x, r)$ is the jth budget constraint, and b_j is the jth available budget, which are related with system weight, volume, and cost.

If all the parameters related to RRAP are fuzzy valued, then Equation (3.1) becomes

TABLE 3.1

Notations Used in the Entire Chapter

x	$(x_1, x_2, \ldots, x_n)$ is the vector of the redundancy allocation of the system
$\tilde{r}$	$(\tilde{r}_1, \tilde{r}_2, \ldots, \tilde{r}_n)$ is the vector of the component reliabilities of the system
n	The number of subsystems
$(x, \tilde{r})$	The vector of the redundancy allocation and component reliabilities of the system
$\tilde{R}_s$	The system reliability
$\tilde{g}_i$	The ith constraint
$\tilde{b}_i$	The upper limit of the ith resource
$\tilde{w}_i$	The weight of the ith subsystem
$\tilde{v}_i$	The volume of the ith subsystem
$\tilde{c}_i$	The cost of the ith subsystem
$\tilde{W}$	The upper bound on the weight
$\tilde{V}$	The upper bound on the volume
$\tilde{C}$	The upper bound on the cost
$\mathbb{Z}^+$	Set of nonnegative integers
$U(a, b)$	Uniform distribution between a and b

$$\text{Maximize } \tilde{R}_S = f(x, \tilde{r}) \qquad (3.2)$$

subject to $\tilde{g}_j(x, \tilde{r}) \leq \tilde{b}_j, \; j = 1, 2, \ldots, m$

$(x, \tilde{r}) = (x_1, x_2, \ldots, x_n, \tilde{r}_1, \tilde{r}_2, \ldots, \tilde{r}_n), \; 1 \leq l_i \leq x_i \leq u_i$ and $x_i \in \mathbb{Z}^+, \; 0 < \tilde{r}_i < 1$ and $i = 1, 2, \ldots, n$.

where $\tilde{R}_S$ is the fuzzy system reliability, $\tilde{g}_j(x, \tilde{r})$ is the jth fuzzy budget constraint, and $\tilde{b}_j$ is the jth available fuzzy budget.

Now, using the ranking of fuzzy number, the problem reduces to

$$\text{Maximize } \mu_{\tilde{R}_S} = f(x, \mu_{\tilde{r}}) \qquad (3.3)$$

subject to $\mu_{\tilde{g}_j(x, \tilde{r})} \leq \mu_{\tilde{b}_j}, \quad j = 1, 2, \ldots, m$

$(x, \mu_{\tilde{r}}) = (x_1, x_2, \ldots, x_n, \mu_{\tilde{r}_1}, \mu_{\tilde{r}_2}, \ldots, \mu_{\tilde{r}_n}), \; 1 \leq l_i \leq x_i \leq u_i$ and $x_i \in \mathbb{Z}^+, \; 0 < \mu_{\tilde{r}} < 1$ and $i = 1, 2, \ldots, n$.

where $\mu_{\tilde{R}_S}$ is the defuzzified system reliability, $\mu_{\tilde{g}_j(x, \tilde{r})}$ is the jth defuzzified budget constraint, and $\mu_{\tilde{b}_j}$ is the jth available defuzzified budget. We now aim at determining the number of components and the component's reliability to maximize the overall system reliability. The problem (3.3) is a mixed-integer constrained optimization problem.

3.5 Constraint-Handling Technique for Constrained Mixed-Integer Nonlinear Problems

The problem (3.3) is a mixed-integer nonlinear constrained optimization problem. For solving the problem (3.3), we have transformed the constrained optimization problem into an unconstrained optimization problem with the help of Big-M penalty function method [18]. Using this method, the corresponding unconstrained optimization problem is as follows:

$$\text{Maximize } \mu_{f(x, \tilde{r})} = \begin{cases} \mu_{f(x, \tilde{r})} & \text{if } (x, \mu_{\tilde{r}}) \in \Omega \\ -M & \text{if } (x, \mu_{\tilde{r}}) \notin \Omega \end{cases} \qquad (3.4)$$

where $\Omega = \left\{ (x, \mu_{\tilde{r}}) : \mu_{g_i(x, \tilde{r})} \leq \mu_{\tilde{b}_i}, \quad i = 1, 2, \ldots, m \right\}$ is the feasible space for the optimization problem.

For minimization problem, $+M$ is considered instead of $-M$. In this work, we have taken the value of M as 99,999.

3.6 Solution Methodology

For solving the problem (3.4), we have developed a hybrid algorithm by assembling two well-known algorithms, namely, GA and PSO-Co. This newly hybridized algorithm has been named as GA–PSO-Co.

In, GA–PSO-Co, we have employed GA for 50% chromosomes from the entire population and PSO-Co for the rest of the chromosomes.

The developed GA–PSO-Co algorithm is given as follows:

Algorithm GA–PSO-Co

Step 1: Set population size (2PS), that is, set the swarm size double by copying the initial population size (PS), the maximum number of generations (MG), the probability of crossover (PC), the probability of mutation (PM), and the bounds of decision variables.

Step 2: Set generation $t = 0$.

Step 3: Initialize the chromosomes/particles of the population $P_{GA}^{(t)}$ and $P_{PSO\text{-}Co}^{(t)}$ as follows:

$$P_{GA}^{(t)} = \left\{ S_k^{(t)}, \quad k = 1, 2, \ldots, PS, \text{ where } S_k^{(t)} = (x, r)_k^{(t)} \right\}$$

$$P_{PSO\text{-}Co}^{(t)} = \left\{ S_k^{(t)}, \quad k = PS+1, PS+2, \ldots, 2PS, \text{ where } S_k^{(t)} = (x, r)_k^{(t)} \right\}$$

Step 4: Find the fitness of $S_k^{(t)}$, $P_{GA}^{(t)}$, and $P_{PSO\text{-}Co}^{(t)}$.

Step 5: Obtain the global best chromosome (S_g) having the best fitness value.

Step 6: Perform the following until the stopping criterion is satisfied:

    i. Apply GA for the entire population $P_{GA}^{(t)}$.

    ii. Find the best chromosome (S_k') from the current population $P_{GA}^{(t)}$.

    iii. Compare S_k' with earlier best chromosome S_k and store better one in S_g.

    iv. Set $t = t + 1$.

    v. Applying selection process to rectify the population of the next generation $P_{GA}^{(t)}$ from previous population $P_{GA}^{(t-1)}$ of $(t-1)$ generation.

    vi. Apply PSO-Co for $P_{PSO\text{-}Co}^{(t)}$.

    vii. Improve the best position of each particle by comparing the position of all chromosomes of $P_{GA}^{(t)}$.

    viii. Calculate the velocity of each particle.

    ix. Obtain the new position of each particle.

   x. Improve the position of each particle and also find the global best particle (S_g).

Step 7: Print the position and fitness of global best particle.

Step 8: End.

3.6.1 Genetic Algorithm

The fundamental idea of GA [34] is to clone the natural evolution artificially in which populations go through uninterrupted changes via genetic operators, such as crossover, mutation, and selection. Distinctly, GA has been found to be more effective in solving complicated real-world optimization problems that cannot be solved simply by direct or gradient-based mathematical techniques. For details and discussions about GA, one may refer to the work of Gen and Cheng [35].

Here we shall discuss concisely about three main operators of GA, namely, selection operator, crossover, and mutation.

In this GA, real coding has been used for representing the chromosomes. Each chromosome/solution has $2n$ number of genes, of which the first n genes correspond to integer variables and the rest correspond to floating point variables.

The selection operator plays a major responsibility in GA. The aim of this operator is to sort out the above average solutions and remove the below average solutions from the population for the next generation by reflecting the principle "survival of the relatively fit." In this work, a tournament selection process of size two has been used with replacement as the selection operator.

After the selection process, a crossover operator is used for the resulting chromosomes that have survived. In this work, we have used intermediate crossover [21] for integer variables. For floating point variables, we have used a power crossover scheme.

The different steps of power crossover are as follows:

Step 1: Calculate integral value of [PC.PS] and keep it in number of classes (NC).

Step 2: Select $s_k^{(t)}$ and $s_i^{(t)}$ randomly from the population at t-th generation.

Step 3: Generate $g = U\left(0, \left|s_{kj}^{(t)} - s_{ij}^{(t)}\right|\right)$, $j = 1, 2, \ldots, n$ and $\lambda = U(0,1)$

Step 4: Calculate $\bar{s}_{kj}^{(t)}$ and $\bar{s}_{ij}^{(t)}$ ($j = 1, 2, \ldots, n$) of two offspring from the parent chromosomes $s_k^{(t)}$ and $s_i^{(t)}$ as follows:

$$\bar{s}_{kj}^{(t)} = \left(s_{kj}^{(t)}\right)^{\lambda} \left(s_{ij}^{(t)}\right)^{1-\lambda} \text{ and } \bar{s}_{ij}^{(t)} = \left(s_{kj}^{(t)}\right)^{1-\lambda} \left(s_{ij}^{(t)}\right)^{\lambda}$$

Step 5: Compute $s_k^{(t+1)} = $ argument of best of $\left\{f(s_k^{(t)}), f(s_i^{(t)}), f(\bar{s}_k^{(t)}), f(\bar{s}_i^{(t)})\right\}$

and $s_i^{(t+1)} = $ argument of best of $\left\{f\left(s_k^{(t)}\right), f\left(s_i^{(t)}\right), f\left(\bar{s}_k^{(t)}\right), f\left(\bar{s}_i^{(t)}\right)\right\}$

Step 6: Perform Steps 2 and 4 for NC/2 times.

Mutation operator assists the random changes into the population used to preclude the search process from converging to the local optima. This operator is applied to a single chromosome only. Here, we have used one-neighborhood mutation for integer variables [19] and nonuniform mutation for floating point/continuous variables.

3.6.2 Particle Swarm Optimization

PSO was initially evolved in the year 1995 [36,37]. It is a population-oriented heuristic method deployed for social interactions and individual experiences. In PSO, each representative is treated as a particle and each particle is a possible solution/result of a prescribed optimization problem. A randomized velocity is generated for every particle, where they pass through the search space of the problem.

Each particle $i(1 \leq i \leq \text{PS})$ has the following characteristics:

    i. A current position $\tilde{x}_i = (\tilde{x}_{i1}, \tilde{x}_{i2}, \ldots, \tilde{x}_{i,2n})$ in the search spaces

    ii. A current velocity $\tilde{v}_i = (\tilde{v}_{i1}, \tilde{v}_{i2}, \ldots, \tilde{v}_{i,2n})$

    iii. A personal best (pbest) position $p_i = (p_{i1}, p_{i2}, \ldots, p_{i,2n})$

It is to be noted that in $\tilde{x}_i$, $\tilde{v}_i$, and p_i, the first n components are integer variables, whereas the rest are floating point variables.

The velocity of each particle is revised as follows:

$$\tilde{v}_{ij}^{(k+1)} = \tilde{w}\tilde{v}_{ij}^{(k)} + \tilde{c}_1 \rho_{1j}^{(k)} \left(p_{ij}^{(k)} - \tilde{x}_{ij}^{(k)} \right) + \tilde{c}_2 \rho_{2j}^{(k)} \left(p_{gj}^{(k)} - \tilde{x}_{ij}^{(k)} \right),$$

$$j = 1, 2, \ldots, 2n; k = 1, 2, \ldots, \text{MG}$$

i.e., $\qquad \tilde{v}_i^{(k+1)} = \tilde{w}\tilde{v}_i^{(k)} + \tilde{c}_1 \rho_1^{(k)} \left(p_i^{(k)} - \tilde{x}_i^{(k)} \right) + \tilde{c}_2 \rho_2^{(k)} \left(p_g^{(k)} - \tilde{x}_i^{(k)} \right) \qquad (3.5)$

In the preceding equation, $\tilde{v}_{ij}^{(k)}$ is the jth component velocity of the ith particle in the kth iteration, $\tilde{w}$ is the inertia weight, $\tilde{c}_1$ and $\tilde{c}_2$ are the acceleration coefficients, and $\rho_{1j}^{(k)}$, $\rho_{2j}^{(k)}$ are two random numbers, where $\rho_{1j}^{(k)} = U(0,1)$ and $\rho_{2j}^{(k)} = U(0,1)$.

The new position of the ith particle is quantified as

$$\tilde{x}_{ij}^{(k+1)} = \tilde{x}_{ij}^{(k)} + \tilde{v}_{ij}^{(k+1)} \text{ i.e., } \tilde{x}_i^{(k+1)} = \tilde{x}_i^{(k)} + \tilde{v}_i^{(k+1)} \qquad (3.6)$$

If the objective function of the optimization problem is maximized, then the personal best (pbest) position of each particle is reequipped as follows:

$$S_i^{(0)} = \tilde{x}_i^{(0)}$$

$$S_i^{(k+1)} = \begin{cases} S_i^{(k)} & \text{if } f\left(\tilde{x}_i^{(k+1)}\right) \leq f\left(S_i^{(k)}\right) \\ \tilde{x}_i^{(k)} & \text{if } f\left(\tilde{x}_i^{(k+1)}\right) > f\left(S_i^{(k)}\right) \end{cases} \tag{3.7}$$

The global best (gbest) position found by any particle during all previous iterations S_g is defined as

$$S_g^{(k+1)} = \arg \max_{S_i} f\left(S_i^{(k+1)}\right), \quad 1 \leq i \leq PS \tag{3.8}$$

3.6.2.1 Constriction Coefficient–Based PSO

Clerc [38] and Clerc and Kennedy [39] recommended the constriction factor and defined it by $\chi = 1 / \left| 2 - \phi - \sqrt{\phi^2 - 4\phi} \right|$. According to Clerc and Kennedy [37], the updated velocity is as follows:

$$\tilde{v}_i^{(k+1)} = \chi \left[\tilde{v}_i^{(k)} + \tilde{c}_1 \rho_1^{(k)} \left(p_i^{(k)} - \tilde{x}_i^{(k)} \right) + \tilde{c}_2 \rho_2^{(k)} \left(p_g^{(k)} - \tilde{x}_i^{(k)} \right) \right]$$

where $\phi = \tilde{c}_1 + \tilde{c}_2$, $\phi > 4$. Usually, $\tilde{c}_1$ and $\tilde{c}_2$ are both set to be 2.05. Thus, ϕ is set to 4.1 and the value of χ is 0.727. This PSO is known as PSO-Co.

In the calculation of $\tilde{v}_i^{(k+1)}$, the rounding off integral value $\left[\tilde{v}_i^{(k+1)} \right]$ is regarded as it corresponds to the first 50% components of each particle.

3.7 Numerical Examples

In this section, we have taken a complicated/complex (bridge) system in the fuzzy environment (see Figure 3.1). The corresponding reliability function of this system is given as follows:

$$\tilde{f}(x, \tilde{r}) = \tilde{R}_1 \tilde{R}_2 + \tilde{R}_3 \tilde{R}_4 + \tilde{R}_1 \tilde{R}_4 \tilde{R}_5 - \tilde{R}_1 \tilde{R}_2 \tilde{R}_3 \tilde{R}_4 - \tilde{R}_1 \tilde{R}_2 \tilde{R}_3 \tilde{R}_5 - \tilde{R}_1 \tilde{R}_3 \tilde{R}_4 \tilde{R}_5 - \tilde{R}_1 \tilde{R}_2 \tilde{R}_4 \tilde{R}_5$$

$$+ 2\tilde{R}_1 \tilde{R}_2 \tilde{R}_3 \tilde{R}_4 \tilde{R}_5$$

Subsequently, the optimization problem is given as follows:
Problem 3.1

$$\text{Maximize } \tilde{f}(x, \tilde{r})$$

subject to

$$\tilde{g}_1(x, \tilde{r}) = \sum_{i=1}^{5} \tilde{v}_i x_i^2 \leq \tilde{V}$$

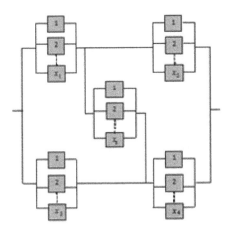

FIGURE 3.1
A complex bridge system.

$$\tilde{g}_2(x,\tilde{r}) = \sum_{i=1}^{5} \tilde{\alpha}_i \left(\frac{-1000}{\ln \tilde{r}_i}\right)^{\tilde{\beta}_i} \left(x_i + \exp\left(\frac{x_i}{4}\right)\right) \leq \tilde{C} \qquad (3.9)$$

$$\tilde{g}_3(x,\tilde{r}) = \sum_{i=1}^{5} \tilde{w}_i x_i \exp(\frac{x_i}{4}) \leq \tilde{W}$$

$$\text{and } \tilde{R}_i = \tilde{R}_i(x_i, \tilde{r}_i) = 1 - (1 - \tilde{r}_i)^{x_i}$$

where $x_i \in \mathbb{Z}^+$ and $0 < \tilde{r}_i < 1$. The parameters $\tilde{\alpha}_i$ and $\tilde{\beta}_i$ are physical features of system components, which are fuzzy valued. The input parameters are given in Table 3.2.

TABLE 3.2

Input Parameters for the Complex System (Problem 3.1)

Stage	$10^5 \tilde{\alpha}$	$\tilde{\beta}_i$	$\tilde{v}_i$	$\tilde{w}_i$	$\tilde{V}$	$\tilde{C}$	$\tilde{W}$
1	(2.33, 2.35, 2.36)	(1.4, 1.5, 1.6)	(0.5, 1, 1.5)	(6, 7, 8)	(108, 110, 113)	(170, 175, 178)	(198, 200, 205)
2	(1.45, 1.46, 1.47)	(1.4, 1.5, 1.6)	(1, 2, 3.5)	(7, 8, 8.5)			
3	(0.541, 0.542, 0.543)	(1.4, 1.5, 1.6)	(2, 3, 4)	(7.8, 8, 8.5)			
4	(8.048, 8.050, 8.051)	(1.4, 1.5, 1.6)	(3.5, 4, 5)	(5, 6, 7.5)			
5	(1.948, 1.950, 1.953)	(1.4, 1.5, 1.6)	(1, 2, 3)	(8.6, 9, 9.5)			

Now, using defuzzification, the above problem (3.9) reduces to

$$\text{Maximize } \mu_{\tilde{f}(x,\tilde{r})}$$

subject to

$$\sum_{i=1}^{5} \mu_{\tilde{v}_i} x_i^2 \le \mu_{\tilde{V}}$$

$$\sum_{i=1}^{5} \mu_{\tilde{\alpha}_i} \left(\frac{-1000}{\ln Y(\tilde{r}_i)} \right)^{\mu_{\tilde{\beta}_i}} \left(x_i + \exp\left(\frac{x_i}{4} \right) \right) \le \mu_{\tilde{C}} \qquad (3.10)$$

$$\sum_{i=1}^{5} \mu_{\tilde{w}_i} x_i \exp\left(\frac{x_i}{4} \right) \le \mu_{\tilde{W}}$$

$$\text{and } \mu_{\tilde{R}_i} = 1 - (1 - \mu_{\tilde{r}_i})^{x_i}$$

where $x_i \in \mathbb{Z}^+$ and $0 < \mu_{\tilde{r}_i} < 1$. The parameters $\mu_{\tilde{\alpha}_i}$ and $\mu_{\tilde{\beta}_i}$ are defuzzified physical features of system components. The defuzzified input parameters are given in Table 3.3.

Here, we have considered RRAP of complicated system for the purpose of numerical experiments. The proposed algorithm has been coded in C++ programming language and corresponding computational work has been done on a PC with Intel Core-2 duo processor in LINUX environment. For computational purpose, the values of parameters such as PS, MG, PC, and PM have been taken as 100, 200, 0.90, and 0.10, respectively. Throughout the computation, 50 independent runs have been carried out for obtaining the best found system reliability, which is only the optimal value of system reliability and the optimal number of redundant components as well as component reliability along with the maximum value of system reliability and the slack of each constraint. Here, the slack of each constraint is the difference between the available resource and the used resource for the best found solution. These results have been shown in Table 3.4.

TABLE 3.3

Input Parameters (Defuzzified Value) for the Complex System (Problem 3.1)

Stage	$\mu_{10^5\tilde{\alpha}}$	$\mu_{\tilde{\beta}_i}$	$\mu_{\tilde{v}_i}$	$\mu_{\tilde{w}_i}$	$\mu_{\tilde{V}}$	$\mu_{\tilde{C}}$	$\mu_{\tilde{W}}$
1	2.347	1.500	1.000	7.000	110.333	174.333	201.000
2	1.460	1.500	2.167	7.833			
3	0.542	1.500	3.000	8.100			
4	8.050	1.500	4.167	6.670			
5	1.950	1.500	2.000	9.000			

TABLE 3.4

Computational Result

$f(x,r)$	$(x_1, x_2, x_3, x_4, x_5, r_1, r_2, r_3, r_4, r_5)$	Slack of the First Constraint	Slack of the Second Constraint	Slack of the Third Constraint
0.99988964	(3, 3, 2, 4, 1, 0.828134, 0.857831, 0.914192, 0.648069, 0.704476)	5.0	0	1.560466

3.8 Concluding Remarks

We have proposed a hybrid approach called GA–PSO-Co by using GA for 50% chromosomes and PSO-Co for the rest. In each chromosome of GA, the first 50% genes relate to integer variables and the remaining 50% genes relate to floating point variables. Here, power crossover operator has been used for floating point variables. For the improvement of the first 50% genes of each chromosomes, intermediate crossover and one-neighborhood mutation have been applied, whereas for the remaining genes, power crossover and uniform mutation have been used. In each and every iteration of PSO, the particle best position is considered by comparing the population of GA. On the other hand, the global best particle of PSO-Co is generated by comparing both populations. For further research, one can refine the proposed hybrid approach using advanced genetic operator, namely, crossover and mutation for GA and velocity upgradation for PSO-Co. Finally, it can be mentioned that the proposed approach employed in this chapter is capable of solving mixed-integer nonlinear programming problems in different areas of science and engineering in the years to come.

References

1. M.S. Chern, On the computational complexity of reliability allocation in a series system, *Operations Research Letters* 11(5) (1992) 309–315.
2. Y. Nakagawa, K. Nakashima, A heuristic method for determining optimal reliability allocation, *IEEE Transactions on Reliability* R-26(3) (1977) 156–161.
3. J.H. Kim, B.J. Yum, A heuristic method for solving redundancy optimization problems in complex systems, *IEEE Transactions on Reliability* R-42(4) (1993) 572–578.
4. W. Kuo, C.L. Hwang, F.A. Tillman, A note on heuristic methods in optimal system reliability, *IEEE Transactions on Reliability* 27(5) (1978) 320–324.
5. K.K. Aggarwal, J.S. Gupta, Penalty function approach in heuristic algorithms for constrained, *IEEE Transactions on Reliability* 54 (2005) 549–558.

6. C. Ha, W. Kuo, Multi-path heuristic for redundancy allocation: the tree heuristic, *IEEE Transactions on Reliability* 55 (2006) 37–43.
7. C.L. Hwang, F.A. Tillman, W. Kuo, Reliability optimization by generalized Lagrangian-function and reduced-gradient methods, *IEEE Transactions on Reliability* R-28 (1979) 316–319.
8. W. Kuo, H. Lin, Z. Xu, W. Zang, Reliability optimization with the Lagrange multiplier and branch-and-bound technique, *IEEE Transactions on Reliability* R-36 (1987) 624–630.
9. F.A. Tillman, C.L. Hwang, W. Kuo, Determining component reliability and redundancy for optimum system reliability, *IEEE Transactions on Reliability* R-26(3) (1977) 162–165.
10. X.L. Sun, D. Li, Optimality condition and branch and bound algorithm for constrained redundancy optimization in series system, *Optimization and Engineering* 3 (2002) 53–65.
11. C.S. Sung, Y.K. Cho, Branch-and-Bound redundancy optimization for a series system with multiple-choice constraints, *IEEE Transactions on Reliability* 48 (1999) 108–117.
12. K.B. Misra, U. Sharma, An efficient algorithm to solve integer programming problems arising in system reliability design, *IEEE Transactions on Reliability* 40 (1991) 81–91.
13. Y. Nakagawa, S. Miyazaki, Surrogate constraints algorithm for reliability optimization problems with two constraints, *IEEE Transactions on Reliability* R-30 (1981) 175–180.
14. M. Hikita, Y. Nakagawa, H. Narihisa, Reliability optimization of systems by a surrogate constraints algorithm, *IEEE Transactions on Reliability* 41(3) (1992) 473–480.
15. B. Dengiz, F. Altiparmak, A.E. Smith, Local search genetic algorithm for optimal design of reliable networks, *IEEE Transactions on Evolutionary Computation* 1(3) (1997) 179–188.
16. R. Tavakkoli-Moghaddama, J. Safar, F. Sassani, Reliability optimization of series-parallel systems with a choice of redundancy strategies using a genetic algorithm, *Reliability Engineering and System Safety* 93 (2008) 550–556.
17. Z. Ye, Z. Li, X. Min, Some improvements on adaptive genetic algorithms for reliability-related applications, *Reliability Engineering and System Safety* 95 (2010) 120–126.
18. R.K. Gupta, A.K. Bhunia, D. Roy, A GA based penalty function technique for solving constrained redundancy allocation problem of series system with interval valued reliabilities of components, *Journal of Computational and Applied Mathematics* 232 (2009) 275–284.
19. A.K. Bhunia, L. Sahoo, D. Roy, Reliability stochastic optimization for a series system with interval component reliability via genetic algorithm, *Applied Mathematics and Computations* 216 (2010) 929–939.
20. A.K. Bhunia, L. Sahoo, Genetic algorithm based reliability optimization in interval environment, Innovative Computing Methods and Their Applications to Engineering Problems, N. Nedjah (Ed.). SCI 357 (2011) 13–36.
21. L. Sahoo, A.K. Bhunia, P.K. Kapur, Genetic algorithm based multi-objective reliability optimization in interval environment, *Computers and Industrial Engineering* 62 (2012) 152–160.

22. L. Sahoo, A.K. Bhunia, D. Roy, An application of genetic algorithm in solving reliability optimization problem under interval component Weibull parameters, *Mexican Journal of Operations Research* 1(1) (2012) 2–19.

23. S.K. Mahato, L. Sahoo, A.K. Bhunia, Reliability-redundancy optimization problem with interval valued reliabilities of components via genetic algorithm, *Journal of Information and Computing Science* 7(4) (2012) 284–295.

24. L. Sahoo, A.K. Bhunia, D. Roy, Reliability optimization in stochastic domain via genetic algorithm, *International Journal of Quality and Reliability Management* 31(6) (2013).

25. L. Sahoo, A.K. Bhunia, D. Roy, Reliability optimization with high and low level redundancies in interval environment via Genetic Algorithm, *International Journal of Systems Assurance Engineering and Management* 5(4) (2014) 513–523. DOI:10.1007/s13198-013-0199-9.

26. Y.C. Hsieh, T.C. Chen, D.L. Bricker, Genetic algorithms for reliability design problems, *Microelectronics Reliability* 38 (1998) 1599–1605.

27. W. Kuo, V.R. Prasad, F.A. Tillman, C.L. Hwang, *Optimal reliability design: fundamentals and applications.* Cambridge: Cambridge University Press (2001).

28. H.G. Kim, C. Bae, D.J. Park, Reliability-redundancy optimization using simulated annealing algorithms, *Journal of Quality in Maintenance Engineering* 12(4) (2006) 354–363.

29. T. Yokota, M. Gen, Y.X. Li, Genetic algorithm for non-linear mixed-integer programming problems and its applications, *Computers and Industrial Engineering* 30(4) (1996) 905–917.

30. L.D.S. Coelho, An efficient particle swarm approach for mixed-integer programming in reliability-redundancy optimization applications, *Reliability Engineering and System Safety* 94 (2009) 830–837.

31. L.A. Zadeh, Fuzzy Sets, *Information and Control* 8(3) (1965) 338–352.

32. R.R. Yager, A procedure for ordering fuzzy subsets of the unit interval, *Information Sciences* 24 (1981) 143–161.

33. A. Rahmani, F. Hosseinzadeh Lotfi, M. Rostamy-Malkhalifeh, T. Allahviranloo, A new method for defuzzification and ranking of fuzzy numbers based on the statistical beta distribution, *Advances in Fuzzy Systems* 2016 (2016) 1–8.

34. D.E. Goldberg, *Genetic Algorithms in Search, Optimization, and Machine Learning, Reading.* Boston, MA: Addison-Wesley Publishing Company, Inc. (1989).

35. M. Gen, R. Cheng, *Genetic Algorithms and Engineering Design* (1st ed.). New York: John Wiley & Sons (1997).

36. R.C. Eberhart, J. Kennedy, A new optimizer using particle swarm theory, *Proceeding of the 6th International Symposium on Micro Machine and Human Science* (1995) 39–43.

37. J. Kennedy, R.C. Eberhart, Particle swarm optimization, *Proceeding of the IEEE International Conference on Neural Networks*, Perth, Australia 4 (1995) 1942–1948.

38. M. Clerc, The swarm and queen: towards a deterministic and adaptive particle swarm optimization, *Procedings of IEEE Congress on Evolutionary Computation*, Washington, DC, USA 3 (1999) 1951–1957.

39. M. Clerc, J. Kennedy, The particle swarm- explosion, stability and convergence in a multidimensional complex space, *IEEE Transactions on Evolutionary Computation* 6(1) (2002) 58–73.

4

Optimal Software Testing Effort Expending Problems

Shinji Inoue
Kansai University

Shigeru Yamada
Tottori University

CONTENTS

4.1 Introduction 51
4.2 Two-Dimensional Software Reliability Model 52
    4.2.1 Basic Assumptions 53
    4.2.2 Reliability Assessment Measures 53
4.3 Optimal Testing Effort Expending Problems 55
4.4 Optimal Policies 57
    4.4.1 Optimal Policy 1 58
    4.4.2 Optimal Policy 2 58
4.5 Numerical Examples 59
4.6 Concluding Remarks 62
References 63

4.1 Introduction

Software reliability measurement/assessment is an interesting topic for developing dependable software systems. A software reliability growth model (Musa et al. 1987; Pham 2000; Yamada 2014) is a useful mathematical model for software project management for developing quality software. After measuring and assessing software reliability, we have other interesting issues on software testing management, such as problems of how long we need to continue testing activities and how much we need to expend testing effort in terms of software development/maintenance cost and software reliability objective. A software project manager has to conduct testing activities economically and efficiently with limited testing resources due to the limited delivery, cost, and testing effort. Therefore, estimating the shipping time of a software product and testing effort expenditure supports to develop a management plan for software testing. At this time, the cost spent throughout the testing phase and the operational maintenance must be a main criterion for estimating the shipping time and the amount of the testing effort expenditure. Generally, it is known that there exists a trade-off relationship between the debugging cost in testing and operation phases (Yamada and Osaki 1985).

In particular, rather than focusing on when to stop the testing activities, the software project manager might be more interested in the testing effort required during the fixed duration of the testing phase and also the appropriate amount of testing effort from the view point of quality and economic efficiency. This is because the delivery schedule of a software system is pre-decided at the contracting session, and the software project manager has to plan a software development schedule for completing the testing activities in the testing phase. This chapter discusses a problem on estimating optimal testing effort expenditures minimizing the expected total software cost, which consists of costs spent in the testing and operational phases, based on a two-dimensional software reliability growth model (Inoue and Yamada 2008, 2009). These models describe the reliability growth process depending on both the types of software reliability growth factors. For example, Ishii et al. (2006) proposed a modeling framework of a two-dimensional software reliability growth model, where the process depends on calendar time and the number of executed test cases. Inoue and Yamada (2008) proposed a Weibull-type model by applying the Cobb–Douglas type function to the reliability growth factors in the existing univariate model. Inoue and Yamada (2009) also proposed a bivariate model with effect of the program size based on a two-dimensional binomial process. Further, in this chapter, we will derive optimal testing effort expending policies in terms of the expected total software cost and software reliability objective. Finally, we will demonstrate numerical examples of the application of derived optimal testing effort expending polices by using actual data.

4.2 Two-Dimensional Software Reliability Model

The reliability growth process is influenced by execution time, testing skill, testing coverage, and so forth (Yamada et al. 1986; Fujiwara and Yamada 2001; Inoue and Yamada 2004). Further, as the related works that handle multiple testing factors, such as execution time, skill, and coverage, Shibata et al. (2006) and Okamura et al. (2010) discussed how to handle multi-factors observed in a testing process in reliability assessment. One of the simplest approaches for visually showing the relationship of testing time and other growth factors, two-dimensional modeling approaches have been proposed

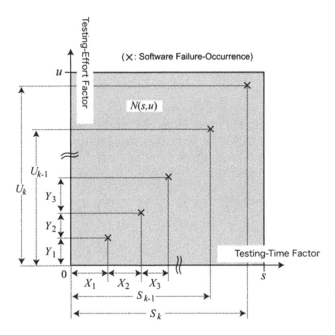

FIGURE 4.1
Two-dimensional software failure-occurrence phenomenon.

so far. As shown in Figure 4.1, the stochastic quantities are described for both the types of reliability growth factors (Inoue and Yamada 2009).

4.2.1 Basic Assumptions

Most of the two-dimensional software reliability growth models can be developed by following basic assumptions (Inoue and Yamada 2009). The basic assumptions of this modeling are summarized in that the software failure-occurrence time distribution follows $G(s,u) \equiv \Pr\{S \le s, U \le u\}$ and the initial fault content is treated as the nonnegative random variable N_0. The two-dimensional probability mass function of $Z(s,u)$ is given as

$$\Pr\{Z(s,u) = m\} = \sum_n \binom{n}{m} \{G(s,u)\}^m \{1 - G(s,u)\}^{n-m} \Pr\{N_0 = n\} \quad (m = 0,1,2,\ldots),$$

(4.1)

where $Z(s,u)$ is the number of faults detected in $(0,s] \times (0,u]$.

4.2.2 Reliability Assessment Measures

Quantitative measures for software reliability assessment can be derived from Equation (4.1). For example, the expectation of $Z(s,u)$ is given as

$$E[Z(s,u)] = \sum_{z=0}^{n} z \sum_{n} \binom{n}{z} \{G(s,u)\}^{z} \{1 - G(s,u)\}^{n-z} \Pr\{N_0 = n\}$$

$$= E[N_0]Z(s,u).$$

(4.2)

We can derive an operational software reliability. The software reliability is a measure for obtaining the probability of failure-free operation in the time interval $(s_e, s_e + \eta](s_e \geq 0, \eta \geq 0)$ given that the amount of the testing effort expenditure has been taking up to u_e by testing time s_e (Ishii et al. 2006). Therefore, this measure is formulated as

$$R_S(\eta \mid s_e, u_e) = \sum_{k} \Pr\{Z(s_e + \eta, u_e) = k \mid Z(s_e, u_e) = k\} \Pr\{Z(s_e, u_e) = k\}$$

$$= \sum_{k} \left[\{G(s_e, u_e)\}^{k} \{1 - G(s_e + \eta, u_e)\}^{-k} \sum_{n} \binom{n}{k} \{1 - G(s_e + \eta, u_e)\}^{n} \right].$$

$$\Pr\{N_0 = n\}$$

(4.3)

We call Equation (4.3) a testing time-dependent operational software reliability function in this chapter. On the other hand, we can propose another software reliability function, in which we assume that the effort influencing the software reliability growth process, such as testing coverage, is expended in the operational phase, and the effect of the operation time factor can be ignored compared with the effort factor. Figure 4.2 shows the basic concept of the testing effort-dependent operational software reliability. The testing effort-dependent software reliability function can be derived as

$$R_U(\eta \mid s_e, u_e) = \sum_{k} \Pr\{Z(s_e, u_e + \eta) = k \mid Z(s_e, u_e) = k\} \Pr\{Z(s_e, u_e) = k\}$$

$$= \sum_{k} \left[\{G(s_e, u_e)\}^{k} \{1 - G(s_e, u_e + \eta)\}^{-k} \sum_{n} \binom{n}{k} \{1 - G(s_e, u_e + \eta)\}^{n} \right].$$

$$\Pr\{N_0 = n\}$$

(4.4)

Further, we can derive three types of measures on the expected instantaneous number of detected faults. For example, the expected instantaneous number of faults at $(\bar{s}, u)$ is given by

$$h_U(\bar{s}, u) = \frac{\partial E[Z(\bar{s}, u)]}{\partial u}$$

(4.5)

from Equation (4.2).

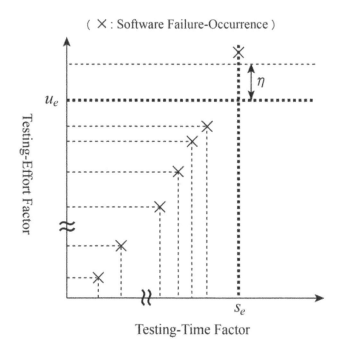

FIGURE 4.2
Testing effort-depended operational software reliability.

4.3 Optimal Testing Effort Expending Problems

We discuss optimal expenditure of testing effort that minimizes the expected software cost consisting of software testing and maintenance costs, under the given testing period and the software reliability objective. Figure 4.3 shows an illustration on a software failure-occurrence pattern in our optimal testing effort expending problems. The following cost parameters are used for describing the trade-off relationship mentioned before under the given testing period:

c_1: the eliminating cost per fault discovered in the testing

c_2: the eliminating cost per fault discovered in the operation ($c_1 < c_2$)

c_3: the running cost expended per simultaneous time and effort

By using these cost parameters, the testing cost is formulated as $c_1 E\left[Z(\bar{S}, U) \right] + c_3 \bar{S} U$ and the maintenance cost as $c_2 \left(E\left[Z(\bar{S}, \infty) \right] - E\left[Z(\bar{S}, U) \right] \right)$, where we define U as the effort expenditure expended up to the fixed testing termination $\bar{S}$. The expected cost function under the given testing period is

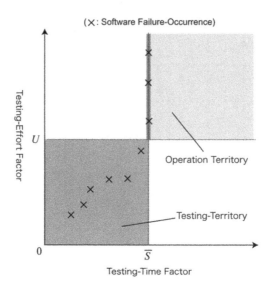

FIGURE 4.3
Software failure occurrence pattern.

$$C(\overline{S},U) = c_1 E\left[Z(\overline{S},U)\right] + c_2\left(E\left[Z(\overline{S},\infty)\right] - E\left[Z(\overline{S},U)\right]\right) + c_3 \overline{S} U. \qquad (4.6)$$

From Equation (4.6), the optimal testing effort expenditure based on the cost criterion minimizes $C(\overline{S},U)$ in Equation (4.6) and satisfies the following equation:

$$\frac{\partial C(\overline{S},U)}{\partial U} = (c_2 - c_1)\left[\frac{c_3\overline{S}}{c_2 - c_1} - h_U(\overline{S},U)\right] = 0. \qquad (4.7)$$

Further, when we consider both the cost and the software reliability objective for discussing a more realistic problem on software project management, this problem can be extended as

$$\left.\begin{array}{c} \text{Minimize } C(\overline{S},U) \\ \text{subject to } R_U(\eta \,|\, \overline{S},U) \geq R_U^0 \ \overline{S} > 0, \, U \geq 0 \end{array}\right\}, \qquad (4.8)$$

where R_U^0 is the software reliability objective for the testing effort-dependent operational software reliability in Equation (4.4).

4.4 Optimal Policies

We derive optimal testing effort expending policies analytically based on the cost and the simultaneous cost and software reliability criteria. Now, we consider that the following model (Inoue and Yamada 2009) is applied for describing the process. Further, we assume that N_0 in Equation (4.1) follows

$$\Pr\{N_0 = n\} = \binom{K}{n} \lambda^n (1-\lambda)^{K-n} \quad (0 < \lambda < 1; n = 0,1,2,\ldots,K). \tag{4.9}$$

Regarding the physical meaning of Equation (4.9), λ is the probability that a fault is still remaining in each line, K is the total lines of codes of the system (Kimura et al. 1993; Inoue and Yamada 2009). Substituting Equation (4.9) into Equation (4.1), we get

$$\Pr\{Z_B(s,u) = m\} = \binom{K}{m} \{\lambda G(s,u)\}^m \{1 - \lambda G(s,u)\}^{K-m} \quad (m = 0,1,2,\ldots,K). \tag{4.10}$$

Regarding the function of $G(s,u)$, we apply Gumbel-type bivariate probability function (Gumbel 1960):

$$G(s,u) = (1 - e^{-as})(1 - e^{-bu})(1 + qe^{-as-bu})(a > 0,\ b > 0,\ -1 \le q \le 1), \tag{4.11}$$

as one of the examples. Its expectation is obtained as

$$E[Z_B(s,u)] = K\lambda(1 - e^{-as})(1 - e^{-bu})(1 + qe^{-as-bu}). \tag{4.12}$$

Similarly, the testing effort-dependent operational software reliability function defined in Equation (4.4) is derived as

$$R_U(\eta \mid s_e, u_e) = \left[1 - \lambda\{G(s_e, u_e + \eta) - G(s_e, u_e)\} \right]^K. \tag{4.13}$$

And the expected instantaneous number of detected faults are derived as

$$h_U(\bar{s}, u) = K\lambda abq e^{-as-bu} \left\{ 1 + q\left(2e^{-as} - 1\right)\left(2e^{-bu} - 1\right) \right\}, \tag{4.14}$$

from Equation (4.5). We derive the optimal policies based on Equation (4.12), assuming the software failure occurrence time distribution follows Equation (4.11). Regarding the cost-optimal policy, we can see

$$h_{UU}(\bar{S},U) = \frac{\partial h_U(\bar{S},U)}{\partial U} = -K\lambda b^2 e^{-bU}(1-e^{-a\bar{S}})\left[1+qe^{-a\bar{S}(4e^{-bU}-1)}\right](>0), \quad (4.15)$$

and

$$\left\{h_U(\bar{S},0) = K\lambda b(1-e^{-a\bar{S}})\left[1+qe^{-a\bar{S}}\right](>0), h_U(\bar{S},\infty) = 0, \quad\quad (4.16)\right.$$

in Equation (4.14) when $\bar{S} > 0$. From Equations (4.15) and (4.16), we can say that $h_U(\bar{S},U)$ in Equation (4.7) is monotonically decreasing with U. Therefore, there exists a unique solution, $U = U_0$, which minimizes $C(\bar{S},U)$, when $\bar{S} > 0$ and $c_3\bar{S}/(c_2 - c_1) < h_U(\bar{S},0)$. Consequently, the optimal policy considering only cost is discussed in Section 4.4.1.

4.4.1 Optimal Policy 1

1. If $h_U(\bar{S},0) > c_3\bar{S}/(c_2 - c_1)$, then there exists a unique solution $U^* = U_0$ minimizing $C(\bar{S},U)$, and the optimal testing effort expenditure U^* is a solution satisfying

$$e^{-bU^*}\left[1+q(1-A)(2e^{-bU^*}-1)\right] = \frac{c_2\bar{S}}{K\lambda Ab(c_2 - c_1)}, \quad\quad (4.17)$$

   where $A = 1 - e^{-a\bar{S}}$.
2. If $h_U(\bar{S},0) \le c_3\bar{S}/(c_2 - c_1)$, then the optimal testing effort expenditure U^* is estimated as 0.

Further, we derive a cost-reliability-optimal testing effort expending policy by adding the constraint of the testing effort-dependent operational software reliability to the Optimal Policy 1. We can see that $R_U(\eta\,|\,\bar{S},0) = 0$, $R_U(\eta\,|\,\bar{S},\infty) = 1$, and $dR_U(\eta\,|\,\bar{S},U)/dU > 0$ when $\eta(> 0)$ and $\bar{S}(> 0)$ are constants from Equation (4.13). Then the testing effort-dependent operational software reliability $R_U(\eta\,|\,\bar{S},U)$ is a monotonically increasing function of U for the constant η. Therefore, when $R_U(\eta\,|\,\bar{S},0) < R_U^0$, there exists a unique solution U_1 satisfying $R_U(\eta\,|\,\bar{S},0) = R_U^0$. And, when $R_U(\eta\,|\,\bar{S},0) \ge R_U^0$, $R_U(\eta\,|\,\bar{S},U) \ge R_U^0$, there exists a unique solution for any nonnegative real number satisfying $U \ge 0$. From the given analytical discussion, we have the following cost-reliability-optimal testing effort expending policy:

4.4.2 Optimal Policy 2

$c_2 > c_1 > 0, c_3 > 0, 0 < R_U^0 < 1, \bar{S} > 0$, and $U \ge 0$.

1. If $h_U(\bar{S},0) > c_3\bar{S}/(c_2 - c_1)$ and $R_U(\eta\,|\,\bar{S},0) \ge R_U^0$, then $U^* = U_0$.
2. If $h_U(\bar{S},0) > c_3\bar{S}/(c_2 - c_1)$ and $R_U(\eta\,|\,\bar{S},0) < R_U^0$, then $U^* = \max\{U_0, U_1\}$.

3. If $h_U\left(\overline{S},0\right) \le c_3\overline{S}/(c_2 - c_1)$ and $R_U\left(\eta\,|\,\overline{S},0\right) \ge R_U^0$, then $U^* = 0$.

4. If $h_U(\overline{S},0) \le c_3\overline{S}/(c_2 - c_1)$ and $R_U(\eta\,|\,\overline{S},0) < R_U^0$, then $U^* = U_1$.

4.5 Numerical Examples

The data applied in this chapter consists of 19 data pairs: $(s_i, u_i, y_i)(i = 1,2,\ldots,19: s_{19} = 19(\text{weeks}),\ u_{19} = 47.65(\text{CPU hours}),\ y_{19} = 328;$ $K = 1.317 \times 10^6(\text{LOC}))$ (Ohba 1984) with respect to the total number of faults, y_i, detected during a constant interval $(0, \Phi_i]$, where Φ_i represents the elements of the space vector, that is, $\Phi_i = s_i,\ u_i$. We should note that the testing time (weeks) and the test-execution time (CPU hours) are treated as the testing time and the testing effort factors, respectively. We need to estimate λ, a, b, and p by applying the method of maximum likelihood where the likelihood function is denoted as l. The likelihood function is derived as (Inoue and Yamada 2009).

$$l = \prod_{i=1}^{N}\binom{K - y_i}{y_i - y_{i-1}}\left\{z(\Phi_{i-1}, \Phi_i)\right\}^{y_i - y_{i-1}}\left\{1 - z(\Phi_{i-1}, \Phi_i)\right\}^{K - y_i}, \qquad (4.18)$$

consequently by using Bayes' formula and Markov property (Osaki 1992; Ross 1997). In Equation (4.18),

$$z(\Phi_{i-1}, \Phi_i) = \frac{\lambda\left\{G(s_i,u_i) - G(s_{i-1},u_{i-1})\right\}}{1 - \lambda G(s_{i-1},u_{i-1})}, \qquad (4.19)$$

where $G(s_0,u_0) = G(s_0,u) = G(s,u_0) = 0$. Then, we can obtain

$L \equiv \log l$

$$= \log K! - \log\left\{(K - y_N)!\right\} - \sum_{i=1}^{N}\log\left\{(y_i - y_{i-1})!\right\} + y_N\log\lambda$$

$$+ \sum_{i=1}^{N}(y_i - y_{i-1})\log\left\{G(s_i,u_i) - G(s_{i-1},u_{i-1})\right\} + (K - y_N)\log\left\{1 - \lambda G(s_N, u_N)\right\}.$$

$$(4.20)$$

We obtain $\hat{\lambda}, \hat{a}, \hat{b}$, and $\hat{q}$, which are the estimation of the parameters λ, a, b, and q, such as $\hat{\lambda} = 0.284 \times 10^{-3}, \hat{a} = 0.132, \hat{b} = 0.630 \times 10^{-1}$ and $\hat{q} = 0.928$, respectively, by solving Equation (4.20) numerically. Figure 4.4 depicts the behavior of $\hat{E}\left[Z_B(s,u)\right]$.

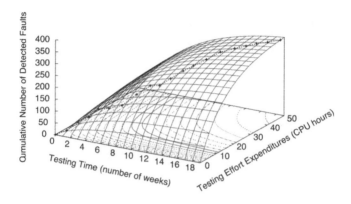

FIGURE 4.4
Estimated expected number of detected faults, $\hat{E}[N_B(s,u)]$.

Now we suppose that $c_1 = 1$ in our discussion for considering the relative-ness for each software cost. Figure 4.5 shows the behavior of $C(\overline{S}, U)$ and $h_U(\overline{S}, U)$, respectively, when $c_1 = 1$, $c_1 = 5$, $c_3 = 1$, and $\overline{S} = 19$. In this case, Optimal Policy (1)-(1) is applied to deriving the cost-optimal testing effort expenditure because $h_U(19,0) \approx 23.2885 > 4.75 = c_3\overline{S}/(c_2 - c_1)$. We obtain the cost-optimal testing effort expenditure for the given testing period $\overline{S} = 19$ as that $U^* = U_0 = 23.4279$. Table 1.1 shows the estimated cost-optimal testing effort expenditures for the several sets of cost parameters. From Table 4.1, we can see that the effort expenditures increase along with increasing the eliminating cost for a fault detected in the operation.

Using the cost-optimal policy discussed with $c_1 = 1$, $c_1 = 5$, $c_3 = 1$, and $\overline{S} = 19$, we analyze an optimal policy with both the cost and reliability crite-ria. We set $\eta = 1.0$ and $R_U^0 = 0.8$. The cost-reliability-optimal problem is then formulated as

$$\left.\begin{array}{c} \text{Minimize } C(19, U) \\ \text{subject to } R_U(1.0 \mid 19, U) \geq 0.8 \; U \geq 0 \end{array}\right\}. \tag{4.21}$$

In this case, Optimal Policy (2)-(2) is applied for getting the cost-reliability-optimal testing effort expenditure because $h_U(\overline{S}, 0) > c_3\overline{S}/(c_2 - c_1)$ and $R_U(1.0 \mid 19,0)(\approx 1.736 \times 10^{-10}) < R_U^0 (= 0.8)$. We can then obtain the cost-reliability-optimal testing effort expenditure as follows: $U^* = \max\{U_0, U_1\} = \max\{23.4279, 70.9731\} = 70.9371$ (see Figure 4.6). From the figure, the software reliability objective $R_U^0 = 0.8$ is not achieved at $U_0 = 23.4279$ under the situa-tion that $\overline{S} = 19$ because $R_U(1.0 \mid 19, 23.4239) \approx 1.007 \times 10^{-2} < 0.8$. Therefore, the manager needs to expend more testing effort to achieve the software reliability objective under $\overline{S} = 19$. From this point of view, we can see the importance of considering both the cost and the software reliability objective as the criteria for obtaining a realistic optimal solution.

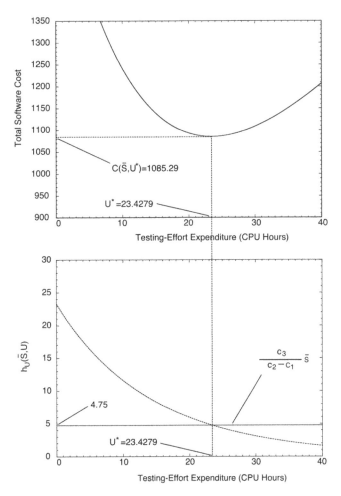

FIGURE 4.5
Behaviors of the total software cost and the expected instantaneous number of faults for $c_1 = 1$, $c_1 = 5, c_3 = 1$, and $\bar{S} = 19$.

TABLE 4.1

Cost-Optimal Testing Effort Expenditures Based on Optimal Policy 1

c_1	c_2	c_3	$h_U(\bar{S},0)$	$c_3\bar{S}/c_2 - c_1$	U^*	$C(\bar{S},U)$
1	2	3	23.2885	57	0	687.763
1	2	2	23.2885	38	0	687.763
1	2	1	23.2885	19	2.86051	681.855
1	3	1	23.2885	9.5	12.9463	881.41
1	4	1	23.2885	6.333	19.0404	1000.28
1	5	1	23.2885	4.75	23.4279	1085.29
1	6	1	23.2885	3.8	26.8598	1151.52

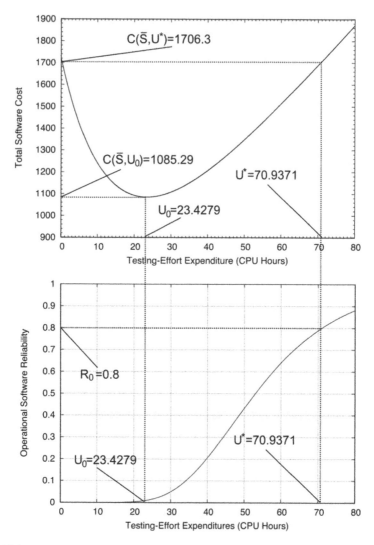

FIGURE 4.6
Comparison of optimal software time for $c_1 = 1$, $c_1 = 5$, $c_3 = 1$, $\bar{S} = 19$, and $R_u^0 = 0.8$.

4.6 Concluding Remarks

We discussed optimal testing effort expending problems under the fixed time duration of testing based on the software cost criterion and the simultaneous cost and operational software reliability criteria, by using the two-dimensional software reliability growth model. Further, we derived optimal testing effort expending policies for obtaining the cost-optimal testing effort expenditure and the cost-reliability-optimal testing effort expenditures analytically, especially for the specific model. This discussion treats different views from existing software development management technologies. A software project manager might have greater interest in the amount of testing effort expended during the fixed testing period for shipping a dependable software system. This is because the delivery of a software system is decided at the contracting session and the software project manager has to plan a software development schedule for completing the testing activities within the fixed delivery of the software system. Further studies are needed to check the effectiveness of our optimal policies by applying them to actual software projects.

References

Fujiwara, T. and S. Yamada. 2001. Software reliability growth modeling based on testing-skill characteristics: model and application. *Electronics and Communications in Japan III* 84:42–49.

Gumbel, E.J. 1960. Bivariate exponential distributions. *Journal of the American Statistical Association* 55:698–707.

Inoue, S. and S. Yamada. 2004. Testing-coverage dependent software reliability growth modeling. *International Journal of Reliability, Quality and Safety Engineering* 11:303–312.

Inoue, S. and S. Yamada. 2008. Two-dimensional software reliability assessment with testing-coverage. In *Proceedings of the Second IEEE International Conference on Secure System Integration and Reliability Improvement*: 150–157.

Inoue, S. and S. Yamada. 2009. Two-dimensional software reliability measurement technologies. In *Proceedings of the 2009 IEEE International Conference on Industrial Engineering and Engineering Management*: 223–227.

Ishii, T., T. Fujiwara and T. Dohi. 2006. Bivariate extension of software reliability modeling with number of test cases. *International Journal of Reliability, Quality and Safety Engineering* 15:1–17.

Kimura, M., S. Yamada, H. Tanaka and S. Osaki. 1993. Software reliability measurement with prior-information on initial fault content. *Transactions of Information Processing Society of Japan* 34:1601–09.

Musa, J.D., D. Iannio and K. Okumoto. 1987. *Software Reliability: Measurement, Prediction, Application*. New York: McGraw-Hill.

Ohba, M. 1984. Software reliability analysis models. *IBM Journal of Research and Development* 28:428–443.

Okamura, H., Y. Etani and T. Dohi. 2010. A multi-factor software reliability model based on logistic regression. In *Proceedings of the 21st International Symposium on Software Reliability Engineering*: 31–40.

Osaki, S. 1992. *Applied Stochastic System Modeling*. Belin-Heidelberg: Springer-Verlag.

Pham, H. 2000. *Software Reliability*. Singapore: Springer-Verlag.

Ross, S.M. 1997. *Introduction to Probability Models* (sixth edition). San Diego: Academic Press.

Shibata, K., K. Rinsaka and T. Dohi. 2006. Metrics-based software reliability models using non-homogeneous Poisson processes. In *Proceedings of the 17th International Symposium on Software Reliability Engineering*: 52–61.

Yamada, S. and S. Osaki. 1985. Cost-reliability optimal release policies for software systems. *IEEE Transactions on Reliability* R-34(5):422–424.

Yamada, S., H. Ohtera and H. Narihisa. 1986. Software reliability growth models with testing-effort. *IEEE Transactions on Reliability* R-35:19–23.

Yamada, S. 2014. *Software Reliability Modeling: Fundamentals and Applications*. Tokyo/ Heiderberg: Springer-Verlag.

5

Revisiting Error Generation and Stochastic Differential Equation-Based Software Reliability Growth Models

Adarsh Anand and Deepika
University of Delhi

A. K. Verma
Western Norway University of Applied Science

Mangey Ram
Graphic Era Deemed to be University

CONTENTS

5.1   Introduction 65
5.2   Methodology 68
      5.2.1   Notations 68
      5.2.2   Assumptions 68
5.3   Data Analysis 72
5.4   Conclusion 75
References 76

5.1 Introduction

The rapid transition of human activities to the computer-based automated system has shown a drastic change in the society. The adaptability of information technology in our daily life has made it a deep-down part of our life. In this virtual world, the technology revolution has made our social system both easier as well as prone to threats. Only considering its easiness, the benefits a user could get through a new innovation changes life to a measurable extent. On the other hand, its failure results in immense loss of human lives and money. With the growth of information and technology, the competition at the ground can give straight benefits to a customer. To capture the potential market, the software firms try to launch their offering at an early stage. Also, due to the increased complexity in design and code, most of software engineers attempt to apply different testing modules comprising various tools and techniques. The increased quality level, in terms of reliability of the software and the features it offers, would be one of the sole criteria that the developer offers to its users.

Software reliability is employed to capture the life and stability of software systems having various criteria. These criteria comprise failure time, testing cost, quality of software, and other attributes. The quality of software incorporates various aspects such as the occurrence of fault, severity of fault, software fault detection rate, and some associated modules. As software testing is a major part of quality control, prior reliability computation is required for the software product delivery. In this context, various software reliability growth models (SRGMs) can be employed for the assessment of reliability. The SRGMs are probabilistic methods for modeling the detection/removal of faults present in the system. They rely on the historical system and fault data, for example, the system usage information, the number of faults detected/ removed, and the effort information. These models can be employed to anticipate different characteristics about the potential faults existing in the system and also when these faults can trigger failures. These models allow the software developer to gauge the amount of testing needed to meet the reliability criteria for the software. The first ever article in software reliability was written by Hudson in the year 1967, where software development was linked with the natural birth and death process (Hudson, 1967). He tried to explain the fact that generation of fault was analogous to birth, whereas any fault corrections were termed as death. After Hudson, one major step that is considered as groundbreaking work in measuring reliability was by Jelinski and Moranda (1972). Several other researches tried to measure software reliability, such as Moranda (1975), Shooman (1972), Schneidewind (1972), and Schick and Wolverton (1973). In 1975, Musa proposed a methodical approach to quantify the software reliability based on execution time. Further, Bayesian approach to measure the software reliability was proposed by Littlewood and Verrall (1973). Goel and Okumoto (1979) proposed an SRGM based on the assumption that the fault causing a failure is immediately removed and is also called as exponential SRGM (commonly termed as G–O model). It is one of the widely used models that exist in nature for modeling real-life scenarios. Yamada et al. (1983) proposed an SRGM to model the concept of failure observation and corresponding fault removal phenomenon, which is also known for its S-shape, in which there is time lag between bugs observed and corrected from the system (Øksendal, 2013). Out of the many S-shaped models proposed, some of them are by Ohba (1984), Bittanti et al. (1988), Kapur and Garg (1992), and so on. Later on, several researchers tried to relax the assumptions of the G–O model, that is, perfect debugging environment, and applied more generic concept that included imperfect debugging, error generation, change point, testing effort, and software release time problem (Kapur et al., 2011b). Researchers have also tried to cater the case of imperfect debugging environment under the idea of prolonged testing in which software firms issue patches to fix failures in operational phase wherein it may happen that some patches turn infectious, that is, lead to increased fault count (Anand et al., 2016). The concept of imperfect debugging and change point is not only concise to single generation but is also extended to the concept of multiple versions of the software (Kapur et al., 2011a).

Thus, it can clearly be seen that a larger number of NHPP-based SRGMs have been studied in the literature of software reliability engineering (SRE). Researchers have discussed various aspects of software growth models under different constraints. Most of the previous researchers have employed the deterministic behavior for the fault detection process. A substantial number of researchers have also studied the impact of uncertain factors suggesting the influence of randomness in the fault detection rate. Randomness means nothing but changes or conditions that affect the testing environment during the debugging process. Generally, randomness arises due to several factors such as testing effort expenditure and testing efficiency and skill, which might influence the testing process. If the size of software system is large, and during the prolonged testing period the amount of bugs observed becomes sufficiently large before the system is released in the market, whereas count of bugs observed and removed throughout the debugging activities resulted sufficiently small as compared to original faults. Hence, the fault removal phenomena can be well described by stochastic process (Goel and Okumoto, 1979). Due to the infusion of various factors involved during the debugging process, such as testing efficiency, efforts, team skills, and strategies, there is an inclusion of noise factor in the fault removal phenomenon that advocates the stochastic behavior of the testing process.

It was Yamada et al. (1994) who first introduced the exponential-based Stochastic Differential Equation (SDE)-based SRGM. The model assumed that fault detection rate is constant along with the noise factor. Later, Yamada et al. (2003) considered different types of fault detection rates to obtain different types of reliability measures. On similar lines, Shyur (2003) captured the behavior of stochastic-based software growth modeling under the influence of imperfect debugging and change point. Lee et al. (2004) employed the stochastic differential equation to illustrate per-fault detection rate that advocates random fluctuation instead of a nonhomogeneous Poisson process (NHPP). Tamura and Yamada (2006) have extended their work and developed a flexible irregular fluctuation model considering distribution development environment. They also discussed an optimum release time considering the reusable rate of software components. Tamura and Yamada (2009) implemented the stochastic-based reliability model to assess the active state of the open source project. They considered the failure intensity as a function of time and the bug tracking system that reports the software faults account an irregular state. Moreover, Kapur et al. (2009) employed logistic error detection rate in the modeling of SDE-based generalized Erlang model. Kapur et al. (2012) also developed a unified approach to formulate various SDE-based SRGMs. Lately, Singh et al. (2009) inculcated the impact of randomness in the formulation of multi up-gradation software releases. Recently, Singh et al. (2015) modeled a multi up-gradation framework that deploys the concept of randomness with learning effect and impact of faults severity.

Accordingly, studying fault removal phenomenon using logistic rate under the influence of imperfect debugging and stochasticity yielded closed-form solution for only one form of time-dependent fault-count structure. To overcome such limitations, in this chapter, an alternative formulation to cater the case of error generation under stochastic environment has been formulated. This mathematical approach not only facilitated the way of obtaining the closed-form solution for the case of pure error generation but also for the case of exponential growth in the fault count and linear form of error generation. Further, the chapter is structured as follows: Section 5.2 illustrates the model formulation for the proposed approach. Section 5.3 presents parameter estimation and criteria used for validation of the proposed model. Section 5.4 provides the conclusion of the chapter.

5.2 Methodology

In this section, emphasis is laid on describing the proposed SDE-based fault removal phenomenon, that too under the case of imperfect debugging environment. Section 5.2.1 lists the set of notations and assumptions that have been used.

5.2.1 Notations

$N(t)$　Population growth function for fault-count phenomenon

$E(N(t)) = m(t)$　Expected number of faults observed or eliminated by time t or mean value function (MVF) of fault removal phenomena

a　Initial amount of fault

$a(t)$　Total fault content dependent on time t

b　Rate of fault removal

β　Learning parameter

α　Constant rate of error generation

σ　Constant representing the magnitude of fluctuations

5.2.2 Assumptions

1. The fault-debugging phenomenon follows the NHPP.
2. The cumulative number of faults being removed by time t is considered as continuous random variable.

3. The rate of fault removal is represented as a stochastic process with continuous state space.

4. The total number of faults is presumed to be an increasing function of time, that is, the fault removal phenomenon is imperfect in nature.

Using the above set of assumptions to capture the uncertainty, we propose an SDE-based fault removal process. Stochastic differential equations have inbuilt tendency to cater randomness and irregular fluctuations that can be represented as ordinary differential equation. This equation describes the Brownian motion, which can be solved by using the Îto integrals. The linear differential equation to model the error growth rate can be described as follows (Kapur et al., 2011a):

$$\frac{dN(t)}{dt} = k(t)\big(a - N(t)\big) \tag{5.1}$$

where $k(t)$ is relative growth rate of errors being detected/removed. Under certain situations, $k(t)$ might not be known in its exact sense; also, it may happen that the fluctuations are because of some environmental factor. In that very case, it is important to account for such randomness that can be modeled by considering associated "noise" term as follows:

$$k(t) = b(t) + \text{"noise"} \tag{5.2}$$

The exact behavior of noise is difficult to understand and only the distribution function that it might follow can be identified in advance, where $k(t)$ is time dependent and nonrandom in nature.

$$k(t) = b(t) + \sigma\lambda(t) \tag{5.3}$$

where $\lambda(t)$ is a factor that portrays the Gaussian white noise and σ accounts for magnitude of irregular fluctuations, that is, $dW(t)/dt = \lambda(t)$. Before we proceed, it is important to understand the basic properties of Wiener process (Brownian motion), which are as follows:

- Wiener process W(t) is a continuous process and $W(0) = 0$.
- Wiener process follows Normal distribution, that is, $W(t) - W(s)$ is $N(0, t - s)$ for $t \geq s \geq 0$.
- Wiener process has independent increments over time.

Equation (5.3) when substituted in Equation (5.1) results in an SDE as follows:

$$\frac{dN(t)}{dt} = \big(b(t) + \sigma\lambda(t)\big)\big(a - N(t)\big) \tag{5.4}$$

Few researchers have considered a fixed count of faults in the software, but in this proposal specifically time-dependent fault content has been considered. So, Equation (5.4) can be rewritten under Itô type as given by (Øksendal, 2013; Yamada et al., 2003)

$$dN(t) = \left(b(t) - \frac{1}{2}\sigma^2 \right)(a(t) - N(t))dt + \sigma\left(a(t) - N(t)\right)dW(t) \qquad (5.5)$$

The complexity of software system influenced the developer to deviate the testing phenomenon from the NHPP-based modeling behavior to a more realistic environment that accounts various uncertain factors during the testing process. The behavior of testing process, that is, majorly stochastic in nature, is influenced by various factors such as testing efforts, testing skills, efficiency, and methods. To capture these uncertain aspects regulating the testing process, various researchers have considered that the behavior of fault detection rate is governed by these factors. Hence, the failure intensity can be modeled that reflect the stochastic property of the testing process considering $b(t)$ as logistic rate, that is, $b(t) = b / (1 + \beta \cdot e^{-bt})$ in Equation (5.5).

Equation (5.5) can be written as

$$dN(t) = \left(\frac{b}{1+\beta \cdot e^{-bt}} - \frac{1}{2}\sigma^2 \right)(a(t) - N(t))dt + \sigma\left(a(t) - N(t)\right)dW(t) \qquad (5.6)$$

On integrating both sides of the differential equation, we get

$$\int dN(t) = \int \left(\frac{b}{1+\beta \cdot e^{-bt}} - \frac{1}{2}\sigma^2 \right)(a(t) - N(t))dt + \int \sigma\left(a(t) - N(t)\right)dW(t) \qquad (5.7)$$

Here, $N(t)$ is considered to be a continuous random variable, and its expected value is given as $E[N(t)] = m(t)$, where $m(t)$ represents the MVF of fault removal process under an uncertain environment. Therefore, on taking expectations on both sides, we get

$$m(t) = \int \left(\frac{b}{1+\beta \cdot e^{-bt}} - \frac{1}{2}\sigma^2 \right)(a(t) - m(t))dt + \int E\left[\sigma\left(a(t) - m(t)\right)dW(t)\right] \qquad (5.8)$$

As per the properties of itô integral, the expectation of second component is zero, which means that the non-anticipating function will be statistically independent in the future of t (Kapur et al., 2005). Hence, the stochastic differential equation can be mathematically modeled as:

$$m(t) = \int \left(\frac{b}{1+\beta \cdot e^{-bt}} - \frac{1}{2}\sigma^2 \right)(a(t) - m(t))dt \qquad (5.9)$$

There are many situations that can portray the concept of error generation. For modeling such scenarios, we consider three different cases of $a(t)$: error generation that is linearly dependent on the amount of error being fixed, exponential growth in error generation, or growth of error that is linearly dependent on time.

Case 1: There are many instances in which testing team might not be able to fix the bugs perfectly; for example, while performing the activity of fixing the bugs some errors were generated leading to increased fault content. These errors were dependent upon the bugs that were getting fixed. So, we have assumed that the amount of error generation is dependent upon the number of bugs being fixed.

On assuming $a(t) = a + \alpha m(t)$ in Equation (5.9) and under the initial condition of $m(t = 0) = 0$, we have the MVF as follows:

$$m(t) = \frac{a}{1-\alpha}\left[1 - \left(\frac{(1+\beta) \cdot e^{-bt}}{1 + \beta \cdot e^{-bt}} \right)^{1-\alpha} \cdot e^{-(1-\alpha)\left(b - \frac{1}{2}\sigma^2 \right)t} \right] \tag{5.10}$$

The preceding equation represents the MVF under stochastic environment based on the case of pure error generation.

When $\sigma \to 0$, Equation (5.10) reduces to the case of logistic model with pure error generation, which was given by Kapur et al. (1999) whose equation can be given as follows:

$$m(t) = \frac{a}{1-\alpha}\left[1 - \left(\frac{(1+\beta) \cdot e^{-bt}}{1 + \beta \cdot e^{-bt}} \right)^{1-\alpha} \cdot e^{-(1-\alpha)b.t} \right] \tag{5.11}$$

Case 2: On the basis of the similar concept of error generation under certain situations, the growth pattern of error generation may follow the exponential pattern, which considers that the initial growth in fault origination is slow, but as the time progresses, there is a tremendous increase in fault count. This is one of the important features of this growth pattern that can be assumed in testing phase.

Consider $a(t) = a \cdot e^{\alpha t}$ in Equation (5.9) and integrating under the initial condition of $m(t = 0) = 0$, then the MVF is defined as follows:

$$m(t) = \frac{a}{1 + \beta \cdot e^{-bt}}\left[\frac{\sigma^2 - 2b}{2\alpha + 2b - \sigma^2}\left(e^{-bt + \frac{1}{2}\sigma^2 t} - e^{\alpha t} \right) + \frac{\beta\sigma^2}{2\alpha - \sigma^2}\left(e^{\frac{1}{2}\sigma^2 t} - e^{\alpha t} \right) \cdot e^{-bt} \right] \tag{5.12}$$

The preceding equation represents the MVF under stochastic environment based on exponentially growth pattern of error generation. In the expression (5.12), if we consider that $\sigma \to 0$, the model converges to the logistic model with exponential growth as given by Kapur et al. (2005):

$$m(t) = \frac{a}{1 + \beta \cdot e^{-bt}} \left[\frac{b}{\alpha + b} (e^{\alpha t} - e^{-bt}) \right]$$

(5.13)

Case 3: Similarly, considering the concept of imperfect debugging under certain situations, the growth pattern of error generation might depend linearly over time, which implies that the errors are being generated in the system with rate α from the initial fault content as the time progresses.

In a similar manner, considering $a(t) = a(1 + \alpha t)$ in Equation (5.9) and integrating under the initial condition of $m(t = 0) = 0$, we have the MVF as follows:

$$m(t) = a \left[1 + \alpha t - \frac{(1 + \beta) \cdot e^{-\frac{1}{2}\sigma^2 t - bt}}{1 + \beta \cdot e^{-bt}} + \frac{2\alpha}{1 + \beta \cdot e^{-bt}} \left(\frac{e^{\frac{1}{2}\sigma^2 t - bt} - 1}{2b - \sigma^2} - \frac{\beta \cdot e^{-bt} \left(e^{\frac{1}{2}\sigma^2 t} - 1 \right)}{\sigma^2} \right) \right]$$

(5.14)

The preceding equation represents the MVF under the stochastic environment based on a linear growth pattern of error generation. When the value of σ is significantly small, the framework tends to a logistic form with linear growth as obtained by Kapur et al. (2011a):

$$m(t) = \frac{a}{1 + \beta \cdot e^{-bt}} \left[(1 - e^{-bt}) \left(1 - \frac{\alpha}{b} \right) + \alpha t \right]$$

(5.15)

Prior knowledge in the field of SRE, the closed-form solution of Cases 2 and 3 were not possible, while in this proposition, we are able to model other two forms of error generation under stochasticity. The literature of SRE is immensely large but there were certain instances which could not result in the closed-form solution or they were quite difficult to be modeled, but the alternative formulations have always proved efficient in capturing such scenarios (Kapur et al., 2011a).

5.3 Data Analysis

For validation of the above proposal, we have considered four real-life software failure data sets. Data Set-I (abbreviated as DS-I) consists of 461 faults that have been removed in 81 weeks from a Brazilian electronic switching system, TROPICO R-1500. Similarly, DS-II, DS-III, and DS-IV consist of 136 bugs removed in 21 weeks, 231 faults removed in 38 weeks, and 481 in

111 weeks, respectively, which are collected from several research articles. (Huang and Hung, 2010; Kanoun et al., 1991; Hwang and Pham, 2008). We have estimated the parameters of model using software package SAS/ETS user's guide version 9.1 (2004) based on nonlinear regression method. To check the performance of the model, we make use of the goodness-of-fit criteria that consist of Sum of Squared Error (SSE), Mean Squared Error (MSE), root mean square, and coefficient of determination (R^2). Tables 5.1–5.4 demonstrate the outcomes of parameter estimation and performance criteria for DS-I, DS-II, DS-III, and DS-IV, respectively.

From Tables 5.3 and 5.4, it can be clearly examined that the values of comparison criteria are quite significant, that is, the obtained values are in the acceptable range as the value of R^2 so obtained is approaching 1, which

TABLE 5.1

Parameter Estimates for DS-I and DS-II

	DS-I			DS-II		
Parameters	Model-I	Model-II	Model-III	Model-I	Model-II	Model-III
a	480.512	468.210	465.560	149.782	148.587	145.403
b	0.119	0.102	0.101	0.416	0.412	0.416
β	0.094	0.139	0.142	570.926	534.700	566.307
α	0.067	0.001	0.001	0.010	0.001	0.002
σ	0.420	0.368	0.365	0.012	0.026	0.011

TABLE 5.2

Parameter Estimates for DS-III and DS-IV

	DS-III			DS-IV		
Parameters	Model-I	Model-II	Model-III	Model-I	Model-II	Model-III
a	299.653	222.020	232.020	484.088	446.939	482.145
b	0.049	0.036	0.034	0.066	0.073	0.064
β	1.120	0.001	0.001	3.647	3.992	1.810
α	0.010	0.011	0.013	0.001	0.001	0.0004
σ	0.012	0.020	0.021	0.001	0.010	0.135

TABLE 5.3

Comparison Criteria for DS-I and DS-II

	DS-I			DS-II		
Comparisons	Model-I	Model-II	Model-III	Model-I	Model-II	Model-III
MSE	86.159	84.035	84.025	7.925	8.154	7.989
SSE	6634.07	6470.036	6469.9	142.60	146.80	143.80
Root MSE	9.282	9.1671	9.1665	2.815	2.855	2.8264
R^2	0.995	0.995	0.995	0.997	0.997	0.997

TABLE 5.4

Comparison Criteria for DS-III and DS-IV

	DS-III			DS-IV		
Comparisons	Model-I	Model-II	Model-III	Model-I	Model-II	Model-III
MSE	29.196	18.606	19.746	300.1	342.2	355.8
SSE	1051.10	669.8	710.9	32409.1	36952.5	38430.7
Root MSE	5.403	4.313	4.444	17.323	18.497	18.864
R^2	0.993	0.995	0.995	0.986	0.985	0.984

means the proposed models are able to cater the fault-removal phenomenon in a good sense. Furthermore, it can also be recognized from Tables 5.3 and 5.4 that the results obtained for Model-II and Model-III are in line with the results of Model-I, which have already existed in the literature. Here we are not comparing these models but proposing an alternative way of obtaining other two forms of error generation. Figures 5.1–5.4 also provide a way to

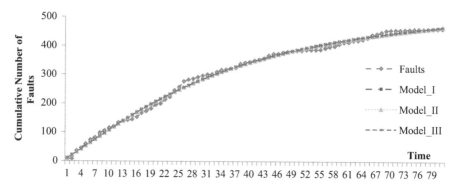

FIGURE 5.1
Goodness-of-fit curve for DS-I.

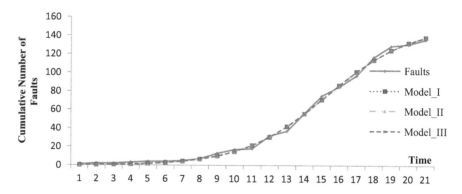

FIGURE 5.2
Goodness-of-fit curve for DS-II.

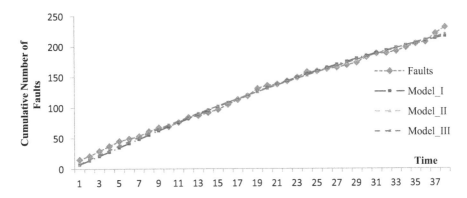

FIGURE 5.3
Goodness-of-fit curve for DS-III.

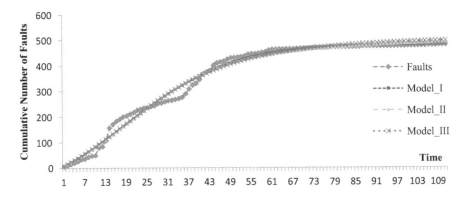

FIGURE 5.4
Goodness-of-fit curve for DS-IV.

graphically analyze the predictive capability of models, which show that computed values are in close correspondence to real-life failure data, thus claiming a fine fitness to considered failure data sets.

5.4 Conclusion

SRGMs are the fundamental technique that can be used to estimate software reliability quantitatively. SRGMs have advocated the software engineer to allocate testing resources and to quantify reliability so as to help the tester in predicting the future behavior of bugs. They are built with the intention to have a good performance in predicting the actual fault count and also to have acceptable values for goodness of fit measures. The major contribution of this work is to capture the impact of uncertainties during the fault-removal process. The proposed SRGM accounts the logistic behavior of fault removal process incorporating the concept of error generation under the influence of randomness. The phenomenon developed here categorizes three possible situations of error generation: pure error generation, linear error generation, and exponential form of error generation. The illustrated methodology has been validated on four real-life software failure data sets and our experiment reveals a better insight of the fault removal process.

References

Anand, A., Das, S., & Singh, O. (2016, September). Modeling software failures and reliability growth based on pre & post release testing. In *2016 5th International Conference on Reliability, Infocom Technologies and Optimization (Trends and Future Directions) (ICRITO)*, pp. 139–144. IEEE.

Bittanti, S., Bolzern, P., Pedrotti, E., Pozzi, M., & Scattolini, R. (1988). A flexible modelling approach for software reliability growth. *Software Reliability Modelling and Identification*, 101–140.

Goel, A. L., & Okumoto, K. (1979). Time-dependent error-detection rate model for software reliability and other performance measures. *IEEE Transactions on Reliability, 28*(3), 206–211.

Huang, C. Y., & Hung, T. Y. (2010). Software reliability analysis and assessment using queueing models with multiple change-points. *Computers and Mathematics with Applications, 60*(7), 2015–2030.

Hudson, G. R. (1967). Program errors as a birth and death process. *System Development Corporation. Report SP-3011, Santa Monica, CA.*

Hwang, S., & Pham, H. (2008). Software reliability model considering time-delay fault removal. *Recent Advances in Reliability and Quality in Design*, 291–307.

Jelinski, Z., & Moranda, P. (1972). Software reliability research. *Statistical Computer Performance Evaluation*, 465–484.

Kanoun, K., de Bastos Martini, M. R., & De Souza, J. M. (1991). A method for software reliability analysis and prediction application to the TROPICO-R switching system. *IEEE Transactions on Software Engineering, 17*(4), 334–344.

Kapur, P. K., & Garg, R. B. (1992). A software reliability growth model for an error-removal phenomenon. *Software Engineering Journal, 7*(4), 291–294.

Kapur, P. K., Garg, R. B., & Kumar, S. (1999). *Contributions to Hardware and Software Reliability* (Vol. 3). World Scientific.

Kapur, P. K., Singh, O., & Gupta, A. (2005). Some modeling peculiarities in software reliability. Proceedings Kapur P.K., Verma A.K. (eds.) *Quality, Reliability and Infocom Technology, Trends and Future Directions*. New Delhi: Narosa Publications Pvt. Ltd., 20–34.

Kapur, P. K., Anand, S., Yamada, S., & Yadavalli, V. S. (2009). Stochastic differential equation-based flexible software reliability growth model. *Mathematical Problems in Engineering, 2009.*

Kapur, P. K., Singh, O., & Singh, J. (2011a). Stochastic differential equation based software reliability growth modeling with change point and two types of imperfect debugging. In *Proceedings of 5th National Conference on Computing for Nation Development, Bharti Vidyapeeth's Institute of Computer Application and Management*, New Delhi, INDIACom, pp. 605–612.

Kapur, P. K., Pham, H., Gupta, A., & Jha, P. C. (2011b). *Software Reliability Assessment with OR Applications*. London: Springer.

Kapur, P. K., Anand, S., Yadav, K., & Singh, J. (2012). A unified scheme for developing software reliability growth models using stochastic differential equations. *International Journal of Operational Research*, 15(1), 48–63.

Lee, C. H., Kim, Y. T., & Park, D. H. (2004). S-shaped software reliability growth models derived from stochastic differential equations. *IIE Transactions*, 36(12), 1193–1199.

Littlewood, B., & Verrall, J. L. (1973). A Bayesian reliability growth model for computer software. *Applied Statistics*, 22(3), 332–346.

Moranda, P. B. (1975). Prediction of software reliability during debugging. In *Proceedings Annual Reliability and Maintainability Symposium*.

Ohba, M. (1984). Software reliability analysis models. *IBM Journal of Research and Development*, 28(4), 428–443.

Øksendal, B. (2013). *Stochastic Differential Equations: An Introduction with Applications*. Springer Science & Business Media.

SAS Institute Inc. (2004). *SAS/ETS User's Guide Version 9.1*. Cary, NC: SAS Institute Inc.

Schick, G. J., & Wolverton, R. W. (1973). Assessment of software reliability. In *Vorträge der Jahrestagung 1972 DGOR/Papers of the Annual Meeting 1972*, pp. 395–422. Physica-Verlag HD.

Schneidewind, N. F. (1972, December). An approach to software reliability prediction and quality control. In *Proceedings of the December 5–7, 1972, Fall Joint Computer Conference, Part II*, pp. 837–847. ACM.

Shooman, M. L. (1972). Probabilistic models for software reliability prediction. *Statistical Computer Performance Evaluation*, 485–502.

Shyur, H. J. (2003). A stochastic software reliability model with imperfect-debugging and change-point. *Journal of Systems and Software*, 66(2), 135–141.

Singh, O., Kapur, P. K., Anand, A., & Singh, J. (2009). Stochastic Differential Equation based Modeling for Multiple Generations of Software. In *Proceedings of Fourth International Conference on Quality, Reliability and Infocom Technology (ICQRIT), Trends and Future Directions, Narosa Publications*, 122–131.

Singh, J., Singh, O., & Kapur, P. K. (2015). Multi up-gradation software reliability growth model with learning effect and severity of faults using SDE. *International Journal of System Assurance Engineering and Management*, 6(1), 18–25.

Tamura, Y., & Yamada, S. (2006). A flexible stochastic differential equation model in distributed development environment. *European Journal of Operational Research*, 168(1), 143–152.

Tamura, Y., & Yamada, S. (2009). Optimisation analysis for reliability assessment based on stochastic differential equation modelling for open source software. *International Journal of Systems Science*, 40(4), 429–438.

Yamada, S., Ohba, M., & Osaki, S. (1983). S-shaped reliability growth modeling for software error detection. *IEEE Transactions on Reliability*, 32(5), 475–484.

Yamada, S., Kimura, M., Tanaka, H., & Osaki, S. (1994). Software reliability measurement and assessment with stochastic differential equations. *IEICE Transactions on Fundamentals of Electronics, Communications and Computer Sciences, 77*(1), 109–116.

Yamada, S., Nishigaki, A., & Kimura, M. (2003). A stochastic differential equation model for software reliability assessment and its goodness-of-fit. *International Journal of Reliability and Applications, 4*(1), 1–11.

6

Repairable System Modeling Using Power Law Process

K. Muralidharan

The Maharaja Sayajirao University of Baroda

CONTENTS

6.1 Introduction 79
6.2 A Counting Process 80
6.3 Power Law Process 81
    6.3.1 Some Distributional Results 83
    6.3.2 Likelihood Estimation and Confidence Interval 84
    6.3.3 Confidence Intervals Based on Future Observations 85
6.4 Bayes Prediction Intervals Based on Future Observations 87
6.5 Hypothesis Tests for Parameters 88
    6.5.1 A Conditional Test for β 88
    6.5.2 Large Sample Test for β Based on Likelihood Function 89
    6.5.3 Test for β Based on Score Function 90
    6.5.4 Test for β in the Presence of Nuisance Parameter θ 91
    6.5.5 A Test Based on Quadratic Form 92
6.6 Conditional Testing and Confidence Interval 93
6.7 Future Reliability Assessment 98
References 99

6.1 Introduction

Consider a complex system under a repair and maintenance program, so that when the system fails, it is repaired and continues in operation. The successive failure time may either increase or decrease due to maintenance. Suppose the failure time follows a Weibull distribution with shape parameter β. Then, if $\beta < 1$, we say the system is improving, and if $\beta > 1$, this would suggest that the system is wearing out and would need a higher level of maintenance. It is also noted that a repair can lead to another repair or failure or a complete shutdown of a system. Suppose the failure is reoccurring and the time between failures is not independent. In such situations the repair strategies have different influences on the system reliability, usually defined as the probability of no failures in time intervals. For example, in the production of a complex item, the time to produce an item would decrease as more items are produced due to increased knowledge and efficiency. Similarly, the time between accidents may increase with improved safety procedures, which would correspond to a decreasing failure rate ($\beta < 1$). The models that have been applied to such systems are called point process models or nonhomogeneous (time-dependent) Poisson process models. Power law process (PLP) or Weibull process is quite often used in developing repairable system models.

6.2 A Counting Process

Consider a point process and let N $(t, t + h)$ denote the number of events occurring in the time interval $(t, t + h]$, where $t \geq 0$ and $h > 0$. If H_t denotes the history of the failure process up to, but not including, t, then the intensity function of a point process is defined as

$$\lambda(t \mid H_t) = \lim_{\Delta t \to 0} \frac{\Pr\left[\text{a failure in the interval } (t, t + h) \mid H_t\right]}{h}$$

$$\cong \lim_{\Delta t \to 0} \frac{\Pr\left[N(t, t + h) - N(t) \geq 1\right]}{h}$$

A counting process $\{N(t), t \geq 0\}$ is said to be a homogeneous Poisson process (HPP) with intensity $\lambda(t)$ if

  i. $N(0) = 0$

  ii. $\{N(t), t \geq 0\}$ has independent increments

  iii. $P[N(t, t + \delta) = 1] = \lambda(t)\delta + 0(\delta)$

  iv. $P[N(t, t + \delta) \geq 2] = 0(\delta)$

If the time between events is neither independent nor identically distributed, then the process is called non-HPP (NHPP). An NHPP is characterized through its cumulative intensity function $\Lambda(t) = \int_0^t \lambda(x)dx$, or by its time derivative $\lambda(t)$, which gives the instantaneous rate of failures. The shape of

$\lambda(t)$—increasing, decreasing, or otherwise—provides information about the time-dependent nature of the reliability of the system. Suppose we observe the system up to time s and let n be the number of failures occurring at times $t_1 < t_2 < \cdots < t_n$, then the likelihood function is given by

$$L(\theta;t) = \prod_{i=1}^{n} \lambda(t_i) \exp\left\{ -\int_0^s \lambda(t)dt \right\} \qquad (6.1)$$

In the remaining sections of this chapter, we discuss the problem of inferences about the parameters of the process and the issues related to prediction that arise when the PLP is used to model reliability growth. Some distributional results and likelihood-based inferences are discussed in Section 6.3. Various methods of confidence interval estimation for the parameters are also studied. The confidence intervals based on future observations or predictive distribution are given in Section 6.4. The estimates are numerically computed and presented in various tables. A detailed trend test and tests for parameter-based inferences are presented in Section 6.5. The conditional confidence interval (CCI) for the ratio of two PLP shape parameters following the rejection of the hypothesis $H_0: \beta_1 = \beta_2$ is presented in Section 6.6. The issues related to reliability estimation are discussed in Section 6.7.

6.3 Power Law Process

Let $T_1 < T_2 < \cdots < T_n$ be the first n occurrence times of a random point process with $T_0 = 0$. Let $N(t)$ be the number of events occurring between 0 and t. This process was first introduced by Duane (1964) and is defined by the intensity function:

$$\lambda(t) = \left(\frac{\beta}{\theta} \right)\left(\frac{t}{\theta} \right)^{\beta-1}, \quad \theta > 0, \beta > 0, t > 0 \qquad (6.2)$$

Such a model is referred to as the PLP. Here, θ and β are, respectively, called the scale and shape parameter of the process. The process described in Equation (6.2) can also be characterized through its mean value function:

$$m(t) = \left(\frac{t}{\theta} \right)^{\beta} \qquad (6.3)$$

The expected rate of occurrence is $d/dt\, m(t)$, which is nothing but Equation (6.2). For $\beta = 1$, the process reduces to a HPP. Otherwise, a PLP provides a model for a system whose reliability changes as it ages. If $\beta > 1$, it models a deteriorating system, and when $\beta < 1$, it provides a model for reliability growth.

The other convenient way of describing this process is through the intensity function:

$$\lambda(t) = \left(\frac{\beta}{\theta}\right) t^{\beta-1}, \quad \theta > 0, \beta > 0, t > 0 \tag{6.4}$$

with the corresponding mean value function $m(t) = (1/\theta)t^\beta$. Since the time to the first failure for a PLP has a Weibull distribution with shape parameter β and characteristic life θ, the model is also called the Weibull process. The other names of this model are Duane model and the Army Materials System Analysis Activity (AMSAA) model. The adequacy of the PLP for a particular data set can be diagnosed graphically using either Duane plots (Duane, 1964) or some modified total time on tests plots (Klefsjo and Kumar, 1992).

Crow (1982), Ascher and Feingold (1984), and Thompson (1988) discuss the applications of this model and provide some inference procedures. Lee and Lee (1978) and Bain and Engelhardt (1980, 1991) have discussed the point estimation and proposed tests for the parameters of the model. A detailed review of this process was given by Rigdon and Basu (1989, 2000). Jani et al. (1997) and Muralidharan (1999) have proposed various tests for the Weibull process. Muralidharan (2002a) has studied reliability inferences of the Weibull process under different realizations. Gaudoin et al. (2003) have proposed the goodness-of-fit test for the Weibull process based on the Duane plot. Bhattacharjee et al. (2004), Attardi and Pulcini (2005), Zhao and Wang (2005), and Gaudoin et al. (2006) have also studied various inferences on the parameters of the abovementioned model. The conditional inference on the PLP model was first explored by Muralidharan et al. (2007), where they have provided interval estimation for the ratio of intensity parameters. One may also see Muralidharan (2008) for a general review of repairable systems and point process models with real-life applications of repairable systems using point process models. In another study, Muralidharan and Chang (2011) have provided large sample tests based on conditional distribution for the shape parameter in PLP. As a potential application of this model, Shah and Muralidharan (2013) have studied the stock price modeling of Indian market under trough and peak situations.

The papers that discuss inferences on PLP based on Bayesian setup are those of Soland (1969), Bar-lev et al. (1992), Calabria and Pulcini (1997), Sen (2002), Pievatolo and Ruggeri (2004), Guida and Pulcini (2006), and Muralidharan and Shah (2006) among others. The generalized version of this process is the modulated power law process introduced by Lakey and Rigdon (1992) and studied thereafter by Black and Rigdon (1996), Muralidharan (2001, 2002b), Ryan (2003), among others.

6.3.1 Some Distributional Results

Let $T_1, T_2, \ldots, T_n$ denote the first n successive times of occurrences of a PLP with $m(t)$ as given in Equation (6.3). If $Z_j = m(T_j)$, for $j = 1, 2, \ldots, n$, then $Z_1 < Z_2 < \cdots < Z_n$ are distributed as the first n successive occurrences of a Poisson process with intensity $\lambda = 1$. Therefore, by the property of Poisson process, $Z_j - Z_{j-1}, j = 1, 2, \ldots, n$; $Z_0 = 0$ are independent exponentially distributed random variables with mean $1/\lambda = 1$. Hence, the joint probability density function of $Z_1, Z_2, \ldots, Z_n$ is given by

$$f(Z_1, Z_2, \ldots, Z_n) = \exp(-Z_n), \quad 0 < Z_1 < Z_2 < \ldots < Z_n \tag{6.5}$$

from which we get the joint probability density function of $T_1, T_2, \ldots, T_n$ as

$$f(t_1, t_2, \ldots, t_n) = \left(\frac{\beta}{\theta}\right)^n \left(\prod_{i=1}^n \frac{t_i}{\theta}\right)^{\beta-1} e^{-(t_n/\theta)^\beta} \tag{6.6}$$

$0 < t_1 < t_2 < \cdots < t_n < \infty, \beta < 0$. However, one can also obtain Equation (6.6) through the use of Equation (6.1) directly. Further, $f(t_1, t_2, \ldots, t_n)/f(u_1, u_2, \ldots, u_n) = k(t, u)$, for every $(\theta, \beta) \in \Omega$, if $t_n = u_n$ and $\sum_{i=1}^{n-1} \log(t_i) = \sum_{i=1}^{n-1} \log(u_i)$, which establishes the fact that $\left(t_n, \sum_{i=1}^{n-1} \log(t_i)\right)$ is a jointly sufficient statistic for (θ, β). For known β, T_n is the complete sufficient statistic for θ. The following densities can be easily derived by suitably integrating Equation (6.6) as follows:

$$f(t_i) = \int_{t_1} \int_{t_2} \int_{t_3} \cdots \int_{t_{i-1}} \int_{t_{i+1}} \cdots \int_{t_n} f(t_1, t_2, \ldots, t_n) dt_1 dt_2 \ldots dt_{i-1} dt_{i+1} \ldots dt_n \tag{6.7}$$

$$= \frac{\beta}{\theta^{i\beta}\Gamma(i)} t_i^{\beta-1} e^{-(t_i/\theta)^\beta}, \quad 0 < t_i < \infty$$

$$f(t_i, t_n) = \int_{t_1} \int_{t_2} \int_{t_3} \cdots \int_{t_{i-1}} \int_{t_{i+1}} \cdots \int_{t_{n-1}} f(t_1, t_2, \ldots, t_n) dt_1 dt_2 \ldots dt_{i-1} dt_{i+1} \ldots dt_{n-1} \tag{6.8}$$

$$= \frac{\beta^2}{\theta^{n\beta}\Gamma(i)\Gamma(n-i)} t_i^{i\beta-1} t_n^{\beta-1} \left(t_n^\beta - t_i^\beta\right)^{n-i-1} e^{-(t_n/\theta)^\beta}, \quad 0 < t_i < t_n < \infty$$

and

$$f(t_n) = \frac{\beta}{\theta^{n\beta}\Gamma(n)} t_n^{n\beta-1} e^{-(t_n/\theta)^\beta}, \quad t_n > 0 \tag{6.9}$$

Using the preceding equation (6.9), we obtain

$$f(t_i|t_n) = \frac{1}{B(i,n-i)}\left(\frac{\beta}{t_n}\right)\left(\frac{t_i}{t_n}\right)^{i\beta-1}\left[1-\left(\frac{t_i}{t_n}\right)^{\beta}\right]^{n-i-1}, \quad 0 < t_i < t_n \quad (6.10)$$

and

$$f(t_1,t_2,\ldots,t_{n-1}|t_n) = \Gamma(n)\left(\frac{\beta}{t_n}\right)^{n-1}\prod_{i=1}^{n}\left(\frac{t_i}{t_n}\right)^{\beta-1}, \quad 0 < t_1 < t_2 < \cdots < t_n \quad (6.11)$$

The following results are now obvious:

Result 6.1: If $u_i = (t_i/t_n)^{\beta}$, then $u_1, u_2, \ldots, u_{n-1}$ are the order statistics from $U(0, 1)$ and $u_n = 1$.

Result 6.2: If $s = \displaystyle\sum_{i=1}^{n-1}\ln(t_n/t_i)$, then $2\beta s$ has $\chi^2_{2(n-1)}$.

6.3.2 Likelihood Estimation and Confidence Interval

Upon using the intensity function in Equation (6.3), we get the likelihood function as

$$L(\theta;t) = \prod_{i=1}^{n}\left(\frac{\beta}{\theta}\right)\left(\frac{t_i}{\theta}\right)^{\beta-1}\exp\left\{-\int_0^s\left(\frac{\beta}{\theta}\right)\left(\frac{t}{\theta}\right)^{\beta-1}dt\right\}$$

which is equivalent to Equation (6.6). The maximum likelihood estimates (MLE) of θ and β are obtained by solving the likelihood equations $\partial\log l/\partial\theta = 0$ and $\partial\log l/\partial\beta = 0$. The estimates are $\hat{\theta} = t_n/n^{1/\beta}$ and $\hat{\beta} = n/\displaystyle\sum_{i=1}^{n-1}\ln(t_n/t_i)$, respectively, for failure truncated data and $\hat{\theta} = T/n^{1/\beta}$ and $\hat{\beta} = n/\displaystyle\sum_{i=1}^{n}\ln(T/t_i)$, respectively, for time truncated data. The following results are now obvious.

Result 6.3: $2n\beta/\hat{\beta} = 2\beta\displaystyle\sum_{i=1}^{n-1}\ln(t_n/t_i) \sim \chi^2_{2(n-1)}$

Result 6.4: $2\left(t_n/\theta\right)^{\beta} \sim \chi^2_{2n}$

A lower $1 - \alpha$ confidence limit for θ is $\theta_L = \hat{\theta}\exp\left(-q_{1-\alpha}\ln n/\sqrt{n}\hat{\beta}\right)$, where $q_{1-\alpha}$ may be obtained from Table 4.18 of Bain and Engelhardt (1980). Similarly, a lower $1 - \alpha$ confidence limit for β is $\beta_L = \hat{\beta}\chi^2_{\alpha}(2n-2)/2n$. The asymptotic confidence interval for the parameters is obtained through the ML estimation. The estimated Fisher information matrix of the model is

$$
I = \begin{bmatrix} I_{\theta\theta} & I_{\theta\beta} \\ I_{\beta\theta} & I_{\beta\beta} \end{bmatrix}
$$

$$
= \begin{bmatrix} E\left(-\dfrac{\partial^2 \ln(L)}{\partial\theta^2}\right) & E\left(-\dfrac{\partial^2 \ln(L)}{\partial\theta\,\partial\beta}\right) \\[3mm] E\left(-\dfrac{\partial^2 \ln(L)}{\partial\beta\,\partial\theta}\right) & E\left(-\dfrac{\partial^2 \ln(L)}{\partial\beta^2}\right) \end{bmatrix}
$$

$$
= \begin{bmatrix} \dfrac{n}{\theta^2} & \dfrac{n}{\theta\beta}\big[\psi(n+1)-\ln\theta\big] \\[3mm] \dfrac{n}{\theta\beta}\big[\psi(n+1)-\ln\theta\big] & \dfrac{n}{\beta^2}+\dfrac{n}{\beta^2}\big[\psi^{(1)}(n+1)+\{\psi(n+1)-\log\theta\}^2\big] \end{bmatrix}
$$

$$(6.12)$$

According to Gaudoin et al. (2006), the first-order approximation to the estimated variance–covariance matrix is

$$
\hat{I} = \begin{bmatrix} \dfrac{1}{n\hat{\theta}^{2\beta}}\Big[1+\big\{\ln\big(n\hat{\theta}^{\hat{\beta}}\big)\big\}^2\Big] & -\dfrac{\hat{\beta}}{n\hat{\theta}}\ln\big(n\hat{\beta}\big) \\[3mm] -\dfrac{\hat{\beta}}{n\hat{\theta}}\ln\big(n\hat{\beta}\big) & \dfrac{\hat{\beta}^2}{n} \end{bmatrix}
$$

$$(6.13)$$

For large n, the quantities $\hat{\theta}-\theta/\sqrt{\mathrm{Var}(\hat{\theta})}$ and $\hat{\beta}-\beta/\sqrt{\mathrm{Var}(\hat{\beta})}$ are asymptotically standard normal variates, and hence asymptotic confidence intervals for θ and β exist. The numerical estimates using simulation study and real-life data are presented in various tables below (table 6.1 and 6.2).

6.3.3 Confidence Intervals Based on Future Observations

Since the events are time dependent, it is natural to see the future behavior of failures based on the last failure and then study the inferences based on parameters and their functions. For this, we obtain the predictive likelihood based on future m-th observations as

$$f\left(t_{n+m}|t_n\right) = \frac{1}{\Gamma m}\left(\frac{\beta}{\theta}\right)\left(\frac{t_{n+m}}{\theta}\right)^{\beta-1}\left[\left(\frac{t_{n+m}}{\theta}\right)^{\beta} - \left(\frac{t_n}{\theta}\right)^{\beta}\right]^{m-1} e^{-\left(\frac{t_{n+m}}{\theta}\right)^{\beta} + \left(\frac{t_n}{\theta}\right)^{\beta}}, \quad t_n < t_{n+m}$$

(6.14)

See the work of Muralidharan et al. (2008) for derivation. Some important results associated with this conditional density are as follows:

Result 6.5: $2(T_{n+m}/\theta)^{\beta}$ follows a chi-square distribution with $2(n + m)$ degrees of freedom.

Result 6.6: $(T_n/T_{n+m})^{\beta}$ has beta distribution with parameter (n, m).

Result 6.7: A lower $(1 - \alpha)\%$ prediction limit for the m-th failure time is $T_L\left(n, m, \alpha\right) = t_n \exp\left[v f_{1-\alpha}/2(n-1)c\hat{\beta}\right]$, where $f_{1-\alpha} = f_{1-\alpha}[v, 2(n-1)]$ is the $(1 - \alpha)$ percentage point of the F distribution, $c = \psi(n+m) - \psi(n)/n[\psi'(n) - \psi'(n+m)]$, $v = 2n[\psi(n+m) - \psi(n)]c$, and ψ is the digamma function.

A Monte Carlo study based on 2,000 failure truncated PLP of different sample sizes has been carried out to evaluate the estimates. The actual values of θ and β assumed in the simulation are 1.3 and 1.2, respectively. Table 6.1 presents the estimates and confidence interval for the parameters based on $T_1, T_2, \ldots, T_n$ (i.e., corresponding to $m = 0$) and on full likelihood (i.e., based on $T_1, T_2, \ldots, T_{n+m}$). The estimates corresponding to $m = 0$ and $m = 1, 2, \ldots$ are comparable in all cases. The future estimates of parameters and their corresponding predictive intervals based on predictive likelihood are presented in Table 6.2. These estimates are also comparable in all cases.

TABLE 6.1

Estimates and Confidence Limits Based on Full Likelihood

n	m	β			θ		
		Estimate	Lower	Upper	Estimate	Lower	Upper
10	0	1.20422	0.57779	1.83065	1.45320	0.48187	2.42454
	1	1.20584	0.57857	1.83311	1.47715	0.51916	2.43515
	2	1.21329	0.58214	1.84444	1.54653	0.62614	2.46693
	3	1.16928	0.56103	1.77754	1.48131	0.51613	2.44648
20	0	1.23865	0.54438	1.69427	1.35451	0.46621	2.24282
	1	1.10149	0.59483	1.50666	1.39453	0.50004	2.28902
	2	1.10852	0.59226	1.51627	1.42741	0.55106	2.30375
	3	1.08376	0.60136	1.48240	1.35761	0.44012	2.27509
30	0	1.05863	0.62553	1.62132	1.31247	0.48404	2.14091
	1	1.06951	0.67878	1.39073	1.32676	0.47527	2.17825
	2	1.05865	0.68205	1.37661	1.28571	0.44159	2.12983
	3	1.06735	0.67944	1.38791	1.32344	0.49941	2.14716

TABLE 6.2

Future Estimates and their Corresponding 10% Predictive Limits

		β			θ		
n	m	Estimate	Lower	Upper	Estimate	Lower	Upper
10	1	1.2536	0.6767	2.0997	1.7107	0.0352	5.3369
	2	1.2686	0.6777	2.2228	1.8403	0.0691	5.5556
	3	1.2697	0.7064	2.2658	1.9036	0.1305	5.4909
20	1	1.1276	0.7668	1.6139	1.8467	0.0443	5.8839
	2	1.1295	0.7868	1.7101	1.7997	0.1065	5.3791
	3	1.1189	0.7106	1.5585	1.7707	0.0965	5.9033
30	1	1.0916	0.8325	1.4651	1.5902	0.1032	4.8957
	2	1.0632	0.6853	1.4193	1.6425	0.0364	5.5365
	3	1.0497	0.7661	1.3814	1.4916	0.0795	4.3254

6.4 Bayes Prediction Intervals Based on Future Observations

As seen in Section 6.3, it is of interest to predict the time of the mth failure and then to predict the time T_{n+m} at which testing will be completed. In such circumstances, a Bayesian method may be desirable as it allows for prior information to be incorporated into the inferential procedure. Let $g(\theta, \beta)$ denote the joint prior density measuring the uncertainty about the parameters (θ, β). Then the joint posterior density is given by

$$\Pi(\theta, \beta | \text{data}) = \frac{g(\theta, \beta) L(\text{data} | \theta, \beta)}{\int_\theta \int_\beta g(\theta, \beta) L() \text{data} | \theta, \beta \, d\beta \, d\theta}$$

Hence, its predictive density function, given the sample information, can be obtained as:

$$h(t_{n+m} | \text{data}) = \int_\theta \int_\beta f(t_{(n+m)} | t_n) \Pi(\theta, \beta | \text{data}) \, d\beta \, d\theta$$

A Bayes prediction interval of a given probability content α is then defined as the interval I such that

$$P(I | \text{data}) = \int h(t_{n+m} | \text{data}) dt_{n+m} = a$$

The predictive intervals based on both informative and non-informative priors. In particular, for $g(\theta, \beta) = (\theta\beta)^{-1}$, the predictive probability density function (pdf) is obtained as

$$h(t_{n+m}) = \frac{n-1}{t_{n+m}} \frac{\sum_{j=1}^{m}(-1)^{j-1}\binom{m-1}{j-1}\left\{\ln\left[\left(t_{n+m}^{n}/u\right)\left(t_{n+m}/t_{n}\right)^{j-1}\right]\right\}^{-n}}{B(n,m)\left[\ln\left(t_{n}^{n}\middle/\prod_{i=1}^{n}t_{i}\right)\right]^{1-n}}$$ (6.15)

For fixed time $t = T$, the mean number of failures $m(t)$ say $M_T = (T/\theta)^{\beta}$ can be viewed as a process parameter and hence an informative prior for M_T may be assumed as Gamma, as:

$$g(M_T) = \frac{1}{\Gamma a} b^a M_T^{a-1} \exp(-bM_T)$$

where a and b may be related to the prior as $a = (\mu/\sigma)^2$ and $b = \mu/\sigma^2$.

6.5 Hypothesis Tests for Parameters

As seen from the process model, the shape parameter β is of more interest in assessing trends and variations in the process. The scale parameter θ is equally important as it signifies the strength of the process. Some of the tests based on parameters are discussed Sections 6.5.1–6.5.5.

6.5.1 A Conditional Test for β

In Section 6.3.1, we had shown that $\left(t_n, \sum_{i=1}^{n-1}\log(t_i)\right)$ is a jointly sufficient statistic for (θ, β). For known β, T_n is the complete sufficient statistic for θ. Then the conditional distribution of $(t_1, t_2, ..., t_{n-1})$ given t_n according to Equation (6.11) is

$$L_c = f(t_1, t_2, ..., t_{n-1}|t_n) = \sqrt{n}\left(\frac{\beta}{t_n}\right)^{n-1}\prod_{i=1}^{n-1}\left(\frac{t_i}{t_n}\right)^{\beta-1}, \quad 0 < t_1 < t_2 < \cdots < t_n \quad (6.16)$$

The density is free from the nuisance parameter θ and is very useful for studying inferences for β. Testing $H_0: \beta = \beta_0$ versus $H_1: \beta > \beta_0$ in the presence of nuisance parameter θ, the test rejects H_0 if

$$-2\sum_{i=1}^{n-1}\log u_i \le c_\alpha \tag{6.17}$$

where c_α is such that $P_{H_0}\left[-2\sum_{i=1}^{n-1}\log u_i \le c_\alpha\right]=\alpha$ and $u_i = (t_i/t_n)^\beta$, $i = 1, 2, \ldots,$ $n-1$. Also, c_α is the αth percentile of χ^2 distribution with $2(n-1)$ degrees of freedom. See the work of Muralidharan (1999) for derivation and other properties.

To test $H_0{:}\beta = \beta_0$ against $H_0{:}\beta \ne \beta_0$, a uniformly most powerful unbiased test is to reject H_0 if $V(x) \notin (c_1, c_2)$, where $V(x)=\sum_{i=1}^{n-1}\log(t_i/t_n)$, and for large n, $c_1 = (1/\beta_0)\chi^2(2(n-1),\alpha/2)$ and $c_2 = (1/\beta_0)\chi^2(2(n-1),(1-\alpha)/2)$. Both these tests are similar to that of Bain and Engelhardt (1980) obtained under an unconditional setup. It is also possible to suggest a large sample test based on the MLE of β as $\hat{\beta}_c \xrightarrow{as} \beta$ and

$$\sqrt{n-1}\left(\frac{\hat{\beta}_c - \beta}{\sqrt{\beta}}\right) \xrightarrow{d} N(0,1), \tag{6.18}$$

where $\hat{\beta}_c = -(n-1)\Big/\sum_{i=1}^{n-1}\log(t_n/t_i)$ which is the conditional likelihood estimate of β of L_c as given in Equation (6.16).

6.5.2 Large Sample Test for β Based on Likelihood Function

To test $H_0: \beta = \beta_0$ against $H_1: \beta \ne \beta_0$, a large sample test based on the likelihood function is to reject H_0, if

$$nI(\beta_0)\left(\beta_0 - \hat{\beta}\right)^2 > \chi^2_{(1,1-\alpha)} \tag{6.19}$$

where $I(\beta) = E\left(-d^2 \log L/d\beta^2\right)$ is the Fisher information function of β.

Again, if $\mu(t)$ is the likelihood ratio (LR) test and $\log\mu(t) = \log L(t,\hat{\beta}) - \log L(t,\beta_0)$, then the LR test is to reject H_0, if $2\log\mu(t) > \chi^2_{(1,1-\alpha)}$. If we expand $\log L(t,\beta_0)$ around $\hat{\beta}$ by Taylor's series and ignoring powers of the order three, we get $2\log\mu(t) = UV + \varepsilon_n$ where, $U = nI(\beta_0)(\beta_0 - \hat{\beta})^2$ and $V = 1/nI(\beta_0)\left(-\partial^2\log L/\partial\beta^2\right)_{\hat{\beta}}$. Further as $n \to \infty$, U is distributed as $\chi^2_{(1)}$, $V \to 1$, and $\varepsilon_n \to 0$.

If we substitute $I(\beta_0)$ from Equation (6.12) and simplifying Equation (6.21), we get

$$\left(\frac{n}{\beta}\right)^2\left[\psi^{(1)}(n+1)+\{\psi(n+1)-\log\theta\}^2\right]\left(\beta_0 - \hat{\beta}\right)^2 > \chi^2_{(1,1-\alpha)} \tag{6.20}$$

It is seen that the preceding equation 0 involves the value of θ. In order to make the test statistic free from the parameter θ, we use the value of $I(\beta)$ obtained through the likelihood L_c as $I_c(\beta) = (n-1)/\beta^2$. Upon substitution in Equation (6.19), the test is to reject H_0, if

$$\frac{n(n-1)}{\beta_0^2}\left(\beta_0 - \hat{\beta}_c\right)^2 > \chi^2_{(1,1-\alpha)} \tag{6.21}$$

Instead of $I_c(\beta_0)$, if we use the estimate of $I_c(\beta)$ evaluated at the estimate $\beta = \hat{\beta}_c$, then the test statistic is to reject H_0, if

$$\frac{n}{(n-1)}\left(\beta_0 - \hat{\beta}_c\right)^2\left[\sum_{i=1}^{n-1}\log\left(\frac{t_n}{t_i}\right)\right]^2 > \chi^2_{(1,1-\alpha)}. \tag{6.22}$$

6.5.3 Test for β Based on Score Function

The score test introduced by Rao (1974) will reject H_0, if

$$\frac{1}{n}\left(\frac{\partial\log L}{\partial\beta}\right)^2 I^{-1}(\beta_0) \geq \chi^2_{(1,1-\alpha)}. \tag{6.23}$$

The unconditional score test based on the likelihood equation (6.1) is then to reject H_0, if

$$\frac{\left[1+\dfrac{\beta}{n}\displaystyle\sum_{i=1}^{n-1}\ln t_i - \dfrac{\theta\beta}{n}t_n^\beta\ln t_n\right]^2}{\left[1+\psi^{(1)}(n+1)-\{\psi(n+1)-\ln\theta\}^2\right]} \geq \chi^2_{(1,1-\alpha)} \tag{6.24}$$

The preceding equation 4 also depends on θ. Based on the conditional distribution in Equation (6.11), we obtain the conditional score function test that rejects H_0, if

$$\frac{n-1}{n}\left[1-\frac{\beta_0}{n-1}\sum_{i=1}^{n-1}\log\left(\frac{t_n}{t_i}\right)\right]^2 \geq \chi^2_{(1,1-\alpha)} \tag{6.25}$$

The test given in Equation (6.25) is independent of θ.

6.5.4 Test for β in the Presence of Nuisance Parameter θ

To propose a large sample test for β in the presence of θ, we use the Fisher information of the process as follows: If $I(\theta,\beta) = \begin{bmatrix} I_{\theta\theta} & I_{\theta\beta} \\ I_{\beta\theta} & I_{\beta\beta} \end{bmatrix}$, then the Fisher information of β in the presence of θ (See Liang 1983) is obtained as

$$I^{\theta}(\beta) = I_{\beta\beta} - I_{\beta\theta} I_{\theta\theta}^{-1} I_{\theta\beta},$$

where $I(\theta, \beta)$ is given in Equation (6.12); $\psi(z)$ and $\psi^{(1)}(z)$ are the digamma and polygamma functions of order 1, respectively. Then the Fisher information of β in the presence of θ is obtained as

$$I^{\theta}(\beta) = \frac{n}{\beta^2 \left[1 + \psi^{(1)}(n+1) \right]} \tag{6.26}$$

Using the preceding equation 6 in Equation (6.21), we get a large sample test that rejects H_0, if

$$\frac{n^2}{\beta_0^2} \left(\beta_0 - \hat{\beta} \right)^2 \left[1 + \psi^{(1)}(n+1) \right]^{-1} > \chi^2_{(1,1-\alpha)} \tag{6.27}$$

where $\hat{\beta}$ is the estimate of β obtained from the likelihood equation (6.6). If we use the estimate of $I^{\theta}(\beta)$ evaluated at the estimate $\beta = \hat{\beta}$, then the test statistic rejects H_0, if

$$\left(\beta_0 - \hat{\beta} \right)^2 \left[1 + \psi^{(1)}(n+1) \right]^{-1} \left[\sum_{i=1}^{n-1} \log(t_n/t_i) \right]^2 > \chi^2_{(1,1-\alpha)} \tag{6.28}$$

For integer n, $\psi^{(1)}(n+1) = (\pi^2/6) - \sum_{j=1}^{n} 1/j^2$, and for large n, $\psi^{(1)}(n+1) = 1/(n+1)$. See the work of Muralidharan and Chang (2011) for details.

Table 6.3 presents the power of the test statistics obtained in Equations (6.19), (6.22), (6.25), and (6.28) for various values of β and samples sizes. As it is seen, most of the tests are computationally simple and some of them are very powerful. We also present the power of the Bain et al. (1985) test for comparison. Throughout, we have assumed $\theta = 10$, $\alpha = 0.05$, and $\beta_0 = 1$. One can easily see that the score test performance is consistent across the tests.

As discussed earlier, the parameter θ shows the strength of the process; it makes sense to make some inferential procedures for scale parameter as well. As seen from the likelihood equation, the estimate of θ depends on the

TABLE 6.3

Power of the Test ($\alpha = 5\%$)

n	Statistics	$\beta = 0.4$	$\beta = 0.8$	$\beta = 1.0$	$\beta = 1.5$	$\beta = 2.0$	$\beta = 3.0$
10	(6.17)	0.2342	0.0718	0.045	0.3576	0.7996	0.8923
	(6.22)	0.1049	0.0526	0.046	0.2782	0.5477	0.7884
	(6.25)	0.4551	0.3278	0.051	0.5668	0.8128	0.9218
	(6.27)	0.3215	0.3059	0.055	0.5812	0.8341	0.8819
	BE test	0.3814	0.3125	0.055	0.5189	0.6129	0.8985
20	(6.17)	0.3145	0.1109	0.044	0.3814	0.8109	0.9123
	(6.22)	0.2122	0.1510	0.048	0.3314	0.5619	0.8013
	(6.25)	0.4814	0.3898	0.052	0.6120	0.8623	0.9416
	(6.27)	0.3516	0.3214	0.053	0.5918	0.8414	0.9014
	BE test	0.4212	0.3590	0.051	0.6108	0.8534	0.9238
30	(6.17)	0.3614	0.2719	0.052	0.4414	0.8516	0.9423
	(6.22)	0.2911	0.2214	0.051	0.3712	0.6912	0.8413
	(6.25)	0.5612	0.4767	0.051	0.6627	0.9102	0.9789
	(6.27)	0.4103	0.4016	0.052	0.6432	0.8613	0.9219
	BE test	0.4925	0.4325	0.051	0.6456	0.8976	0.9465

shape of the distribution. Further, we state some results from the work of Bain and Engelhardt (1980) as follows:

Result 6.8: A size α test of $H_0{:}\theta \leq \theta_0$ against $H_1{:}\theta > \theta_0$ is to reject H_0 if $\sqrt{n}\hat{\beta}\ln(\hat{\theta}/\theta_0)/\ln(n) \geq q_{1-\alpha}$.

Result 6.9: A lower $1 - \alpha$ confidence limit for θ is $\hat{\theta}\exp(-q_{1-\alpha}\ln(n/\sqrt{n}\hat{\beta}))$, where $q_{1-\alpha}$ may be obtained from Table 18 of Bain (1980).

6.5.5 A Test Based on Quadratic Form

Jani et al. (1997) have proposed a conditional test for β keeping θ as a nuisance parameter based on a quadratic form defined as

$$Q = (\underset{\sim}{x} - \mu)' \sum^{-1} (\underset{\sim}{x} - \mu)$$

where $\mu = (\mu_1, \mu_2, \ldots, \mu_n)'$ is the mean of $\underset{\sim}{x}|\underset{\sim}{t}$, $\Sigma = ((\sigma_{ij}))$ is the variance–covariance matrix of $\underset{\sim}{x}|\underset{\sim}{t}$ computed under the null hypothesis, and $\underset{\sim}{t}$ is the sufficient statistic for θ. Since T_n is complete sufficient for β when θ is known, and the conditional distribution of $t_i|t_n$ shown in Equation (6.10) is free from parameter θ, the conditional mean and variance can be obtained using the density in Equation (6.10). Thus, to test $H_0: \beta = \beta_0$ against $H_1: \beta > \beta_0$, the test statistics has the form

$$Q_1 = 2n(n+1)\sum_{i=1}^{n-1}\frac{T_i}{T_n}\left[\frac{T_i}{T_n}-\frac{T_{i+1}}{T_n}\right]+\left(n^2-1\right) \tag{6.29}$$

Under H_0, the mean and variance of Q_1 is respectively obtained as $E(Q_1)=n-1$ and $V(Q_1) = 4n^2(n-1)/(n+2)(n+3)$. The authors have proposed a similar kind of test with transformed variables as $\ln(T_1)$, $\ln(T_2)$, ..., $\ln(T_n)$ instead of T_1, T_2, ..., T_n and has the form

$$Q_2 = \sum_{i=1}^{n-1}\left[1-(n-i)\left\{\ln\left(T_{n-i+1}\right)-\ln\left(T_{n-i}\right)\right\}\right]^2 \tag{6.30}$$

which has $EH_0(Q_1) = n - 1$ and $VH_0(Q_1) = 24(n - 1)$. Further, if $V_i = \ln(T_n) - \ln(T_{n-i})$, $i = 1, 2, ..., n$ then $U_i = (n - i)(V_i - V_{i-1})$ are independent and identically distributed (iid) standard exponential random variables with $E(U_i) = 1$, then the modified form of the test statistic Q_2 is

$$Q_3 = \sum_{i=1}^{n-1}\left[U_i-\bar{U}\right]^2 \tag{6.31}$$

where $\bar{U} = 1/(n-1)\sum_{i=1}^{n-1}U_i$. The mean and variance of this statistic under HPP assumption is $n - 2$ and $2(4n^2 - 15n + 14)/n - 1$, respectively. One may see Jani et al. (1997) for the asymptotic distribution, percentile points, and power of the tests and their comparisons. It has been shown that the test performs well for $\beta < 0.5$ and $\beta \geq 2$.

6.6 Conditional Testing and Confidence Interval

Using the conditional distribution, Muralidharan et al. (2007) proposed the CCI for $\phi = \beta_1/\beta_2$ upon rejection of the test $H_0{:}\beta_1 = \beta_2$ against $H_1{:}\beta_1 \neq \beta_2$. Let $f_{\alpha/2}$ and $f_{1-\alpha/2}$ denote the $100(\alpha/2)$ and $100(1 - \alpha/2)$ percentage points of the F distribution with $[2(n_1 - 1)2(n_2 - 1)]$ degree of freedom, then an α-level critical region of the test $H_0{:}\beta_1 = \beta_2$ against $H_1{:}\beta_1 \neq \beta_2$ is defined by

$$\phi(x) = \begin{cases} 1, & T < f_{\alpha/2} \text{ or } T > f_{(1-\alpha)/2} \\ 0, & \text{otherwise} \end{cases} \tag{6.32}$$

where $T = (n_2 - 1)s_1/(n_1 - 1)s_2$ and α is such that $\Pr_{H_0}\left[T < f_{\alpha/2} \text{ or } T > f_{(1-\alpha)/2}\right] = \alpha$. Since $(\beta_1/\beta_2)T$ has an F distribution with $[2(n_1 - 1), 2(n_2 - 1)]$ degrees of freedom, the usual $100(1 - p_1 - p_2)\%$ unconditional confidence interval (UCI) for ϕ is $(f_{p_1}/T) \le \phi \le (f_{1-p_2}/T)$, where p_1 and p_2 are the lower and upper tail probabilities. For constructing UCI, we need not have a preliminary test. The power of the test say $\beta(\phi)$ is obtained as

$$\beta(\phi) = \Pr_{H_1}\left[T < f_{\alpha/2} \text{ or } T > f_{1-\alpha/2}\right]$$

$$= \Pr_{H_1}\left[T\left(\frac{\beta_1}{\beta_2}\right) < f_{\alpha/2}\left(\frac{\beta_1}{\beta_2}\right) \text{ or } T\left(\frac{\beta_1}{\beta_2}\right) > f_{1-\alpha/2}\left(\frac{\beta_1}{\beta_2}\right)\right] \qquad (6.33)$$

$$= 1 + F\left(\phi f_{\alpha/2}\right) - F\left(\phi f_{1-\alpha/2}\right)$$

The CCI is constructed based on the conditional distribution of T given the power of the test. If $f(t)$ denotes the unconditional pdf of T, then the conditional pdf of T is obtained as

$$f_c(t) = \begin{cases} \dfrac{f(t)}{\beta(\phi)}, & \text{if } t < f_{\alpha/2} \text{ or } t > f_{(1-\alpha)/2} \\ 0, & \text{otherwise} \end{cases} \qquad (6.34)$$

Note that $f_c(t)$ is defined only when $\beta(\phi) > 0$ and for $\alpha > 0$, $f_c(t)$ is always defined. For ϕ other than unity (say 0 or ∞) and α very large (say $\alpha \to 1$), the conditional density converges to the unconditional pdf. In Figure 6.1, we present the graph of $F_c(t)$ for various values of ϕ. It is observed that for $\phi < 1$, $F_c(t)$ is increasing and for $\phi > 1$, $F_c(t)$ is decreasing. We will use this fact to compute the CCI in the following theorem.

Theorem 6.1

Let $\beta_i(\phi) = 1 + F\left(\phi_i^c f_{\alpha/2}\right) - F\left(\phi_i^c f_{1-\alpha/2}\right)$, $i = L, U$. Then

i. If $p_1 < (1/\beta_L(\phi))F\left(\phi_L^c f_{\alpha/2}\right)$ and $1 - p_2 \le 1/\beta_U(\phi)F\left(\phi_U^c f_{\alpha/2}\right)$, then the solutions of the equations

$$p_1 = \frac{1}{\beta_L(\phi)}F\left(T\phi_L^c\right) \qquad (6.35)$$

and

$$1 - p_2 = \frac{1}{\beta_U(\phi)}F\left(T\phi_U^c\right) \qquad (6.36)$$

constitute a $100(1 - p_1 - p_2)\%$ confidence interval for ϕ.

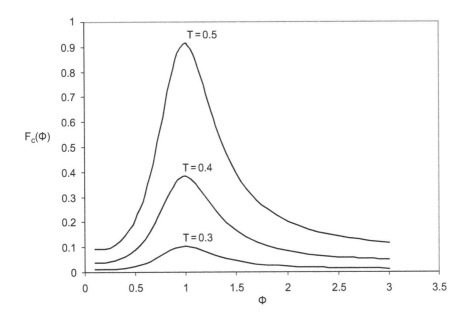

FIGURE 6.1
The graph of $F_c(t)$.

ii. If $p_1 < (1/\beta_L(\phi))F(\phi_L^c f_{\alpha/2})$ and $1 - p_2 > (1/\beta_U(\phi))F(\phi_U^c f_{\alpha/2})$, then the solutions of Equation (6.4) and the following equation

$$1 - p_2 = \frac{1}{\beta_U(\phi)}\left[F\left(T\phi_U^c\right) + F\left(\phi_U^c f_{\alpha/2}\right) - F\left(\phi_U^c f_{1-\alpha/2}\right)\right] \qquad (6.37)$$

constitutes a $100(1 - p_1 - p_2)\%$ confidence interval for ϕ.

iii. If $p_1 \geq (1/\beta_L(\phi))F(\phi_L^c f_{\alpha/2})$ and $1 - p_2 > (1/\beta_U(\phi))F(\phi_U^c f_{\alpha/2})$, then the solutions of the Equation (6.6) and the following equation

$$p_1 = \frac{1}{\beta_L(\phi)}\left[F\left(T\phi_L^c\right) + F\left(\phi_L^c f_{\alpha/2}\right) - F\left(\phi_L^c f_{1-\alpha/2}\right)\right] \qquad (6.38)$$

constitute a $100(1 - p_1 - p_2)\%$ confidence interval for ϕ. The theorem is a direct consequence of results of Equations (6.33) and (6.34).

For given values of T, α, p_1, p_2, the limits ϕ_L^c and ϕ_U^c can be obtained by solving the equations in the above theorem. The ratio of the length of the CCI to that of UCI is given by $R = (\phi_U^c - \phi_L^c)/(f_{1-p_2}/T - f_{p_1}/T)$. Table 6.4 provides the value of the ratio of the length of a 90% CCI to the length of a 90% UCI for $n_1 = 10$, $n_2 = 10$, $\alpha = 0.10$, and for given values of T. Note that the ratio does not exist when $f_{\alpha/2} \leq T \leq f_{1-\alpha/2}$. When T approaches $f_{\alpha/2}$ (=0.45) from the left, the length

TABLE 6.4

Ratio of the Lengths of CCI to UCI

T	R	T	R
0.01	1.801	2.5	0.223
0.05	1.785	3.0	0.478
0.10	1.583	3.5	0.735
0.15	1.405	4.0	0.856
0.20	1.020	4.5	0.923
0.25	0.998	5.0	0.976
0.30	0.983	10.0	1.004
0.35	0.727	20.0	1.063
0.40	0.357	30.0	1.078
0.45	0.035	40.0	1.089

of the CCI can be much smaller than the length of the UCI. As T increases to infinity or decreases to zero, the ratio exceeds one and eventually converges to one from above.

To study the actual coverage probability that is provided at the nominal $100(1 - p)\%$ level, we compute the coverage probability of the UCI under the conditional pdf of T. This is provided in the following theorem.

Theorem 6.2

If $\phi f_{\alpha/2} \le f_{p/2}$ and $\phi f_{1-\alpha/2} \le f_{1-p/2}$, then the (CCP) of ϕ is zero.

    i. If $\phi f_{\alpha/2} \le f_{p/2}$ and $f_{p/2} < \phi f_{1-\alpha/2} < f_{1-p/2}$, then the (CCP) of ϕ is

$$\frac{1}{\beta(\varphi)}\left[\left(1-\frac{p}{2}\right)-F\left(\varphi f_{1-\alpha/2}\right)\right]$$

    ii. If $\phi f_{\alpha/2} \le f_{1-p/2}$ and $\phi f_{1-\alpha/2} \le f_{p/2}$, then the (CCP) of ϕ is $(1/\beta(\varphi))(1-p)$.

    iii. If $f_{p/2} < \phi f_{\alpha/2} < f_{1-p/2}$ and $f_{p/2} < \phi f_{1-\alpha/2} < f_{1-p/2}$, then the (CCP) of ϕ is

$$\frac{1}{\beta(\phi)}\left[(1-p)+F\left(\phi f_{\alpha/2}\right)-F\left(\phi f_{1-\alpha/2}\right)\right]$$

    iv. If $f_{p/2} < \phi f_{\alpha/2} < f_{1-p/2}$ and $\phi f_{1-\alpha/2} \ge f_{1-p/2}$, then the (CCP) of ϕ is

$$\frac{1}{\beta(\varphi)}\left[F\left(\varphi f_{\alpha/2}\right)-\frac{p}{2}\right]$$

Table 6.5 presents the nominal $100(1 - p)\%$ coverage probability of the UCI for $n_1 = 20$, $n_2 = 15$, $p = 0.10$, and for various values of ϕ and α. As an illustration,

TABLE 6.5

Conditional Coverage Probability of UCI

ϕ	$\alpha = 0.05$	$\alpha = 0.1$	$\alpha = 0.2$	$\alpha = 0.4$	$\alpha = 0.5$
0.1	0.907047	0.902673	0.900822	0.900182	0.900034
0.2	0.941472	0.945695	0.933196	0.910766	0.900235
0.3	0.915146	0.930677	0.940227	0.945244	0.923768
0.4	0.859235	0.898994	0.922124	0.932916	0.916576
0.5	0.750834	0.839796	0.886637	0.904265	0.902365
0.6	0.546975	0.731783	0.819266	0.84684	0.895644
0.7	0.184487	0.542631	0.698736	0.78368	0.854278
0.8	0.1213	0.256254	0.513283	0.751567	0.756785
0.9	0.0000	0.1567	0.427775	0.733468	0.789456
1.0	0.0000	0.0000	0.429422	0.732691	0.776567
1.1	0.0000	0.291832	0.481541	0.745145	0.798998
1.2	0.129789	0.481888	0.550707	0.76404	0.865432
1.3	0.35792	0.611659	0.730987	0.784143	0.898654
1.4	0.510417	0.698333	0.790877	0.802716	0.823675
1.5	0.614178	0.757227	0.83099	0.818777	0.843545
1.6	0.686888	0.798463	0.858518	0.877425	0.887875
1.7	0.739403	0.828268	0.877973	0.895353	0.904535
1.8	0.778397	0.850451	0.892135	0.908179	0.900235
1.9	0.808061	0.867393	0.90273	0.917513	0.903467
2.0	0.831107	0.880622	0.910851	0.924422	0.923676
2.1	0.849336	0.891148	0.917209	0.929619	0.954355
2.2	0.86398	0.899661	0.92228	0.933588	0.945345
2.3	0.8759	0.90664	0.926387	0.936662	0.946758
2.4	0.88571	0.912431	0.929758	0.939072	0.953456
2.5	0.893857	0.917283	0.932559	0.940984	0.957643
2.6	0.900675	0.921383	0.934908	0.942518	0.945678
2.7	0.906411	0.924873	0.936896	0.943759	0.943567
2.8	0.911257	0.927858	0.938591	0.944772	0.946786
2.9	0.91536	0.930424	0.940044	0.945606	0.956787
3.0	0.918836	0.932634	0.941298	0.946298	0.967856

we have obtained the CCI estimation for the example based on the data by Klefsjo and Kumar (1992)) on hydraulic systems of load-haul-dump (LHD) machines discussed in section 6.5. PLP models were used for checking the presence of trends in the time between failures of the hydraulic system. According to them, the null hypothesis of equality of two shape parameters is accepted in every case when two different systems are compared. Although their conclusion is same everywhere, they have obtained the value of test statistic using $T = \hat{\beta}_1 / \hat{\beta}_2$ instead of T as defined above. Since the hypothesis is accepted in every case, the CCI cannot be constructed. The estimates and UCI for ϕ when the two shape parameters are tested are presented in Table 6.6.

TABLE 6.6

UCI for Hydraulic Systems of LHD Machines

LHD Machine	Estimate β	Estimate θ	LHD Comparison	Statistic T	UCI
1	1.628	363.9			
3	1.493	408.0	(1 vs 3)	0.9204	(0.6660, 1.7384)
9	1.654	646.6			
11	1.316	231.7	(9 vs 11)	0.7967	(0.7969, 1.9679)
17	1.528	383.2			
20	1.220	253.2	(17 vs 20)	0.7943	(0.7779, 2.0553)

6.7 Future Reliability Assessment

According to Ascher and Fiengold (1984, pp. 24–25), the true reliability of the PLP depends on the entire history of the failure process. Therefore, we define the reliability of the process at time t based on the distribution of T_i as (see also Muralidharan, 2002a, b):

$$R_i(t) = P(T_i > t)$$

$$= \int_t^\infty f(t_i) dt_i \tag{6.39}$$

$$= 1 - \frac{\beta}{\Gamma(i)\theta^{i\beta}} \int_0^t t_i^{i\beta-1} e^{-(t_i/\theta)^\beta} dt_i$$

This reliability will depend on the ith realizations of the process. If $i = n$, then the reliability will be based on the first n realizations and so on. The future reliability based on Equation (6.14) is

$$R_m(t) = P(T_{n+m} > t | T_n = t_n)$$

$$= \int_t^\infty f(t_{n+m} | t_n) dt_{n+m} \tag{6.40}$$

$$= 1 - \frac{\beta}{\theta\Gamma(m)} e^{-(t_n/\theta)^\beta} \int_{t_n}^t \left(\frac{t_{n+m}}{\theta}\right)^{\beta-1} \left[\left(\frac{t_{n+m}}{\theta}\right)^\beta - \left(\frac{t_n}{\theta}\right)^\beta\right]^{m-1} e^{-(t_{n+m}/\theta)^\beta} dt_{n+m}$$

In the case of the preceding equation 40, since $t_n < t \, t_{n+m}$, it is important to fix the value of time t while computing the future reliability as it can influence the reliability of the system very much. We have used the value of t as $(t_n + t_{n+m})/2$ for $m = 1, 2, \ldots$ One can also explore other possibilities for t. For detailed inferences, one may see Muralidharan et al. (2008) and the references contained therein.

References

Ascher, H. and Feingold, H. (1984). *Repairable Systems Reliability: Modelling and Their Causes*, Marcel Dekker, New York.

Attardi, L., Pulcini, G. (2005). A new model for repairable systems with bounded failure intensity, *IEEE Transactions on Reliability*, 54(4), 572–582.

Bain, L.J. and Engelhardt, M.E. (1980). Inferences on the parameters and current system Reliability for a time truncated Weibull process, *Technometrics*, 22, 421–426.

Bain, L.J. and Engelhardt, M.E. (1991). *Statistical Analysis of Reliability and Life-Testing Models*, Marcel Dekker Inc., New York.

Bain, L.J. and Engelhardt, M.E. and Wright, F.T. (1985). Tests for an increasing trend in the intensity of a Poisson process: A power study, *Journal of the American Statistical Association*, 80(390), 419–423.

Bar-Lev, S.K., Lavi, I., and Reiser, B. (1992). Bayesian inference for the power law process, *Annals Institute Statistical Mathematics*, 44, 623–639.

Bhatacharjee, M., Deshpande, J.V. and Naik-Nimbalkar, U.V. (2004). Unconditional tests of goodness of fit for the intensity of time-truncated non homogeneous Poisson processes, *Journal of the American Statistical Association*, 46(3), 330–338.

Black, S.E. and Rigdon, S.E. (1996). Statistical inference for a modulated power law process. *Journal of Quality Technology*, 28(1), 81–90.

Calabria R. and Pulcini, G. (1997). Bayes inference for the the modulated power law process, *Communication in Statistics: Theory and Methods*, 26(10), 2421–2438.

Crow, L.H. (1982). Confidence interval procedures for the Weibull process with applications to reliability growth, *Technometrics*, 24, 67–72.

Duane, J.T. (1964). Learning curve approach to reliability, *IEEE Transactions on Aerospace*, 2, 563–566.

Gaudoin, O., Yang, B. and Xie, M. (2003). A simple goodness-of-fit test for the Parameter for the power-law process based on the Duane plot, *IEEE Transactions on Reliability*, 52(1), 69–74.

Gaudoin, O., Yang, B. and Xie, M. (2006). Confidence intervals for the scale parameter of the power-law process, *Communication in Statistics: Theory and Methods*, 35(8), 1525–1538.

Guida, M. and Pulcini, G. (2006). Bayesian analysis of repairable systems showing a bounded failure intensity, *Reliability Engineering and System Safety*, 91(7), 828–838.

Jani, P.N., Shanubhogue, A. and Muralidharan, K. (1997). Some tests for poisson process and their powers with respect to trend in the intensity, *Communication in Statistics: Theory and Methods*, 26(11), 2641–2688.

Klefsjo, B. and Kumar, U. (1992), Goodness-of-fit tests for the power-law process based on the TTT-plot, *IEEE Transactions on Reliability*, 41, 593–598.

Lakey, M.J. and Rigdon, S.E. (1992). The modulated power law process. *Proceedings of the 45th Annual Quality Congress*, Milwankee, 559–563.

Lee, L., and Lee, K.S. (1978). Some results on inference for the Weibull process, *Technometrics*, 20(1), 41–45.

Liang, K.Y. (1983). On information and ancilarity in the presence of a nuisance Parameter. *Biometrika*, 70, 607–612.Muralidharan, K. (1999). Testing for the shape parameter in Weibull process: A conditional approach, *Statistical Methods*, 1, 36–40.

Muralidharan, K. (2001). On testing of parameters in modulated power law process, *Applied Stochastic Models in Business and Industry*, 17, 331–343.

Muralidharan, K. (2002a). On reliability inferences of *i*-th Weibull process, *Statistical Methods*, 3(2), 50–63.

Muralidharan, K. (2002b). Reliability inferences of Modulated power law process #i, *IEEE Transactions on Reliability*, 51(1), 23–26.

Muralidharan, K. (2008). A review of repairable systems and point process models, *Journal of ProbStat Forum*, 1, 26–49.

Muralidharan, K. and Shah R. (2006). Bayesian analysis of change point in power law process, *Statistical Methods*, 8(2), 164–178.

Muralidharan, K., Chang, K.C. and Chiuo, P. (2007). Conditional interval estimation for the ratio of intensity parameters in power law process, *Journal of Statistical Research*, 41(2), 21–23.

Muralidharan, K., Shah, R. and Dhandhukia, D.H. (2008). Future reliability estimation based on predictive distribution, *Quantitative Techniques and Quality Management*, 5(3), 193–201.

Muralidharan, K. and Chang, K.C. (2011). Some large sample tests based on conditional distribution for the shape parameter in power law process, *Journal of Statistical Theory and Practice*, 5(2), 327–334.

Pievatolo, A. and Ruggeri, F. (2004). Bayesian reliability analysis of complex repairable Systems, *Applied Stochastic Models in Business and Industry*, 20(3), 253–264.

Rao, C.R. (1974). *Linear Statistical inference and its applications*. Wiley.

Rigdon, S.E. and Basu, A.P. (1989). The power law process: a model for the reliability of repairable systems, *Journal of Quality Technology*, 21(4), 251–260.

Rigdon, S.E. and Basu, A.P. (2000). *Statistical Methods for the Reliability of Repairable Systems*, John Wiley, New York.

Ryan, K.J. (2003). Some flexible families of intensities for non-homogeneous Poisson process models and their Bayes inference, *Quality and Reliability Engineering International*, 19(2), 171–181.

Sen, A. (2002). Bayesian estimation and prediction of the intensity of the power-law process, *Journal of Statistical Computation and Simulation*, 72(8), 613–631.

Shah, R. and Muralidharan, K. (2013). Modeling of Indian stock prices using non homogeneous Poisson process with time trends, *IOSR Journal of Business Management*, 13(4), 73–86.

Soland (1969). Bayesian analysis of the Weibull process with unknown scale and shape parameter, *IEEE Transactions on Reliability*, R-18, 181–184.

Thompson, W.A. (1988). *Point Process Models with Applications to Safety and Reliability*, Chapman and Hall, New York.

Yu, J.W., Tian, G.L., and Tang, M.L. (2007). Predictive analyses for non-homogeneous Poisson Process with power law using Bayesian approach, *Computational Statistics and Data Analysis*, 51, 4254–4268.

Zhao, J. and Wang, J.D. (2005). A new goodness-fo-fit test based on the laplace statistic For a large class of NHPP models, *Communication Statistics – Simulation and Computation*, 34(3), 725–736.

7

Reliability and Safety Management of Engineering Systems through the Prism of "Black Swan" Theory

Iosif Aronov
Research Center "International Trade and Integration"

Ljubisa Papic
DQM Research Center

CONTENTS

7.1 Introduction 103
7.2 "I am looking for an honest man" 104
7.3 Safety Culture 105
7.4 Lessons Learnt 106
7.5 Inductive Thinking 107
7.6 Fifty Shades Darker 108
7.7 Conclusion 110
References 111

7.1 Introduction

The "Black Swan" theory has been described by an American economist and stockbroker of Lebanon ancestry, Nassim Nicholas Taleb, in the book entitled "The Black Swan: The Impact of the Highly Improbable" [1]. According to this theory, it is hard to predict situations and rare events that cause significant consequences. In the opinion of Nassim Taleb, it is not possible to describe the processes of the real world from the exclusively one kind of mathematics point of view. Retrospective analysis of past mishaps carries a little information about possible negative events in the future. Bearing in mind that these events happen very rarely, there is a shortage of scenario descriptions for certain accepted methods and standards.

The "Black Swan" theory mostly describes processes that happen in world-wide economics and stock market trading. This chapter analyzes the extent to which the conclusions of the Black Swan theory are applicable in the contemporary safety theory of engineering systems, relying on the "allowed risk" concept [2].

7.2 "I am looking for an honest man"*

As shown in a number of directions (manuals) for safety management, for example, in Ref. [3], a tendency toward increment in specific gravity of negative events emerged. This aforementioned conclusion is illustrated in Figure 7.1, formed according to Federal State Statistics Service [4] data, which shows the change in safety indicator "quantities of incidents at one billion in ton-kilometers" for different modes of transport.

Dynamics of this indicator shows its cyclicality, which greatly confirms the reasonableness of Nassim Taleb's words—about "people, as a rule, make errors in prognosis on one side: they are overestimating their abilities and capabilities. We tend to underestimate the risk if we have been avoiding it successfully in the past."

Why is that? One of the reasons of the aforementioned cyclicality consists of the fact that as soon as the level of risk decreases in a certain time window (measures for its decrease have proven to be effective in that particular time window), experts (people whom safety greatly depends on) presume that the safety issue was solved entirely, and they decrease their efforts in this area of expertise. As a result, a number of incidents start to increase again and "sinusoids" in Figure 7.1 convert into the increase phase, and so on. Consolation is not even given by Nassim Taleb's conclusion with reference to aforementioned phenomena—"… we do not learn…".

Statistics [5] show that regarding a safety assurance, the "weak link" is the human being: influences of people in car traffic lead to accidents in 57% of cases, in railroad traffic approximately 50%, and in air and river transport up to 70% of cases.

Analysis of other data gives an even sadder picture: approximately 83% of air crashes are caused by human factor [6] and approximately 75% of sea accidents are in connection with this factor [7].

That means that the "Black Swan" problem—engineering systems accidents—can be described through Nassim Taleb's book terminology, at least in part that relates to human error.

---

* Expression attributed to ancient Greek philosopher Diogenes of Sinope (400–325 B.C.).

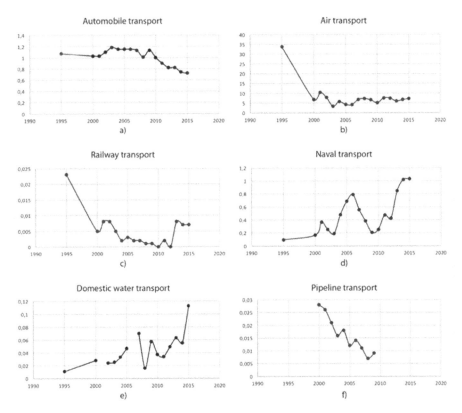

FIGURE 7.1
Dynamics of the transport system safety indicator.

7.3 Safety Culture

What could be confronted to tendency shown in Figure 7.1, which obviously has a global character? First of all, education about safety culture for experts of every level. The safety culture is here observed as the complex of values, world view, behavior habits characteristic to personnel, and as complex of values that affect safety policy, which is applied within an enterprise. In manufacturing activities, the culture manifests itself through

- The thinking of the person who accidentally made an error: the fear of punishment or intention to openly speak about what happened, relying on objective consideration of experts
- Human relationships within the collective: problems are set and settled or are being "run away from"

- Communication among hierarchical structures: listening to an expert in a high position or not listening to him
- The relation toward the expert who made an error: to punish him or to find factors that brought to that error, and so on.

In practice, some dangerous factor is first revealed by a concrete expert in the process of his professional work. Therefore, usually right from the beginning, a timely detection of dangerous factors and operational actions is required for its removal. Besides, a personnel is always a carrier of information about the errors or disturbances made and also about officially non-registered failures and mistakes. Announcement of the given information to the management enables determination of its causes and preparation of effective preventive measures. Certainly, an expert that finds himself in the atmosphere of fear from a punishment for the error or mistake he made will refrain from informing anyone about any of the dangerous factors.

It becomes clear that every worker could become an active participant of the safety assurance system and has to be convinced that the company has created "nonpunitive" working environment in which errors and mistakes are not reasons to be punished. This enables the personnel to build a new relationship toward safety and to understand its part in solving this problem.

However, if the safety culture is being constrained only on reacting to an event, in which case it is obviously not enough for the situation to be improved. For the personnel to actively detect problems that are in relation with safety, it is important to constantly direct them toward continuous searching for risk factors.

We should become aware that corrective measures, which have manifested their effectiveness in some moment in time, could also prove to be ineffective in some other situation. A characteristic example in this regard gives us the analysis of the nuclear power lant "Three Mile Island" (TMI-2) accident (on March 28, 1979, USA). Although operational personnel proceeded as instructed, its actions in the first 8 minutes from the beginning of the accident were wrong because several critical indicators on the control panel were obscured by some extrinsic objects [8]. Obviously, the instructions did not predict such a possibility.

7.4 Lessons Learnt

Here we should make a digression, which relates to "lessons learning" from this accident. Inability of reading the indicator's data indication is a complementary event of the accident and should be excluded in future. However, actions in that direction should be different: if an indicator failure happens, then for preventing the similar event it is necessary to increase its reliability. But what to do in case there is a tape over the indicator (saying: "John, don't forget to take a medicine")?* We are back again on forming the safety culture within the company because only the culture represents *universal* means for preventing personnel from making wrong actions.

Nassim Taleb pays attention on exactly this aspect: one should treat the information obtained from the analysis of rare accidents very cautiously and critically.

7.5 Inductive Thinking

Including standard procedures of new data and new operational algorithms, obtained as a result of previous accidents' analysis, represents a necessary condition for ensuring safety, but it is obviously not enough. Understanding "lessons learning" from accidents caused by object exploitation consists of *not* considering that scenarios for *every possible* accident are already entirely known and thus exploitation instructions should give in detail answers for *every* unpredicted situations. That is why experts responsible for safety do not have a right to be careless and to "rest on their laurels."

In fact, about that, even though in different words, speaks Nassim Nicholas Taleb.

There is another fundamental reason that restricts an expert's possibilities in the safety domain. It is formulated in the form of philosophical principle on inductive opinion restriction on the basis of transition from individual to general situation. An expert meets with this situation using information about the flow of previous situations generalizing it for the current situation. British mathematician and philosopher Bertrand Russell brought to attention the restriction of inductive method thinking (using a hen as an example) [9], and on the very same shortcoming, Nassim Taleb paid attention as well (using a Christmas turkey as an example).

We are tempted to describe another example: who hasn't seen a sleepy keeper in a jewelry shop, who is, as a rule, sleeping or reading a newspaper. Unlike Russel's hen or Taleb's turkey, he can think.† The keeper thinks (evaluates) that in several years there hasn't been any unpredicted situation, so the next time when he comes to his job at the jewelry shop, he puts on his glasses and starts reading the newspaper. Then it results in an attack on the jewelry shop, which ends with stealing of precious jewels. For the keeper, this event is "Black Swan." From the psychologist's point of view, the keeper makes his own opinion, a behavior pattern regarding the rules about duty in

---

* Author's imaginary situation.
† In fact – that is also an example of inductive thinking.

the jewelry shop on the basis of previous situations on the same job, protecting the shop many times before when nothing happened.

From the fact that there were no unpredicted situations for a long period of time, it absolutely does not mean that in the next time window they will not happen at all.

7.6 Fifty Shades Darker*

Inductive thinking, previously spoken about, implicates in fact the entire exploitation system of engineering system for safety. What is that in connection with? If we imagine a potentially dangerous object that is characterized by some N dangerous factors, the realization of each of them during the system exploitation could lead to an accident. In fact, we should consider that in practice these causes could appear per se as well as in different combinations. Total number of these combinations, as we can easily see, is equal to 2^N. And that is already a large number, even if number N is small.

For example, according to Ref. [10], approximately 27 individual causes can lead to accidents on structures and buildings, but the accidents often occur due to unfavorable combination of different failures. In theory, during safety analysis we should consider a possibility of $2^{27}=134217728$ combinations of accident causes, which is unrealistic. For more simple objects, the number of individual accident causes is a lot less: for example, in the paper by Vasiliev and Salnikov [11], seven to nine accident causes of steel reservoirs have been considered, indicating 64–512 combinations in theory, which is also not a small number.

In practice, of course, not all the causes can be manifested at the same time, but even eliminating all impossible combinations of causes leaves a lot of possible accident causes, which should be analyzed.

How is this analysis conducted? First of all, the most probable scenarios of accidental situations are being submitted for analysis. This is clear because it is not possible to analyze all possible accidental situations caused by combinations of causes. The basis of every safety analysis is information about reliability of system elements, which is stored in corresponding databases. If data within the databases are correct, then the results of the analysis can be considered to be valid, but if those data are "poor" or "fake," how can the analysis results be qualified? American physicist and mathematician of Hungarian ancestry, Cornelius Lanczos, conveyed that "no (mathematical) trickery can remedy lack of information."

But what happens with the analysis of highly unlikely accident scenarios? They are waiting for "better" times and the time is coming. As written by

---

* Analogy with the title of James Foley's movie "Fifty Shades Darker."

Professor Boris G. Gordon, analyzing lessons from the Fukushima Daiichi Nuclear Power Plant accident, scenarios of such accidents were unknown earlier [12]. But how is that? Well, because it was assumed that scenarios of these accidents are highly unlikely.

We observe a simple example. Let us evaluate a degree of decrease of the safety reliability system, which consists of two sensors of *the same kind* that transfer the signal to executive devices. Sensors are connected in parallel from the reliability point of view. Data on sensor activation reliability of this kind is in the database.

Let reliability activation of every sensor be P_i, determined according to database, and equal to 0.999, $i=1, 2$. In that case, reliability P_{S1} is determined as

$$P_{S1} = 1 - (1 - P_1)(1 - P_2) = 0.999999.$$

Taking into account high reliability of this device, the scenario of accident caused by failure of this safety system could not be considered.

However, remember that this safety system used two sensors of *the same kind*. That means, under certain conditions, the system could fail due to failure of the sensors and also because of the common cause that is typical for both the products— because of a defect in manufacturing. In that case, calculation of the activating reliability for such a system should consider probability of failure due to common cause. Let us assume that this probability is equal to $Q=10^{-3}$. Activating reliability of the system P_{S2} will be equal to

$$P_{S2} = \left[1 - (1 - P_1)(1 - P_2)\right](1 - Q) = 0.9989.$$

Comparing values P_{S1} and P_{S2}, it can be concluded that because of the failure due to common cause, the system reliability has decreased three times. In earlier scenarios, there was considered a situation exclusive of accident scenario, caused by safety system failure, which was incomplete.

Thereby, inductive thinking of a designer during system design for safety brought negative consequences. The problem was majorly caused by the expert's "belief" in data from database and neglecting the system approach, which in some portion could compensate the inductive thinking.

Whereas scenarios of some accidents "were not known earlier," they were not analyzed in detail in the stage of designing, and its consequences can be proven to be the most sensitive for humans and the environment. Such kind of accidents are called accidents for which no safety systems are predicted during design ("accidents built-in design"), and they are caused by factors that were not taken into consideration during design analysis.

We can say that among "Black Swans" these swans are the blackest ones. Regarding the prognosis of such accidents, we have to agree with the author [12] for one thing: "We can't say where and when, but its scenario will show itself to be impossible." Since those accidents cannot be predicted, the task of the owner of dangerous objects is to assure decrease of its negative impact on the personnel, population, and environment. In that regard, the task of this country is to compel the owners of such objects to strictly carry out directions, regulations, and directives of normative documents in the safety department.

Nassim Taleb believes that humanity has to get used to living in the world of "Black Swans." Accordingly, management issues ("accidents built-in design") are considered in every detail in normative documents of International Atomic Energy Agency (IAEA). In general, management of "accidents built-in design" is directed toward reduction of its consequences, for example, through informing the population, forming the alienation zone, finishing or revising directions (instructions, directives) for personnel, and so on.

7.7 Conclusion

Naturally, there is a question—what is new in the "Black Swan" concept regarding engineering systems safety theory? In Ref. [13] such a question has exactly been asked in accordance with economic science and social phenomenon (events), which were in fact researched in the book of Nassim Taleb.

It is possible to agree with the author [13] that "every individual fragment was multiple discovered long time ago by other researchers..., but interconnected they were turned into original system".

The most important aspect discussed by Nassim Taleb is about the randomness of "Black Swans," the necessity to be prepared for them, and uselessness of its ultimately conscientiously prognosis because of the lack of information. The entire theory "accidents built-in design" of engineering systems actually rejects these postulates.

Thus, there is practically no new thing for technical experts in this doctrine.

Interesting passages of Nassim Taleb are therefore dedicated to the problems of inductive thinking and making decisions. Actually, just that fragmental thinking is frequently manifested by personnel that uses a dangerous object. But corresponding "antidote" was made here as well—forming of safety culture.

As far as the reliability of engineering systems and their complexity increases, the "human factor" part increases as well. Therefore, safety culture education, making of non- polluting manufacturing environment stays the most important task of theory and practice for engineering systems safety.

References

1. Nicholas Taleb N. *The Black Swan: The Impact of the Highly Improbable*. New York: Random House and Penguin. 2007.
2. Haring I. *Risk Analysis and Management: Engineering Resilience*. Springer Singapore+Business Media. 2015. 364 p.
3. Safety Management Manual. Doc 9859. AN/460. ICAO. First Edition. 2006.
4. Official Site of Federal State Statistics Service. www. gks.ru.
5. Smolian G.L., Solntseva G.N. The human factor as a transport security threat. *Transactions of the ISA RAN*. Vol. 64. No. 3. 2014. pp. 81–90.
6. Popov Yu.V. Safety performance of aviation flights. *Internet-Journal "Technologies of Technospheric Safety"*. Vol. 6. No. 58. 2014. http://ipb.mos.ru//ttb.
7. Skorokhodov D.A., Borisova L.F., Borisov Z.D. Principles and categories of ensuring the safety of navigation. *Bulletin of MSTU*. Vol. 13. No. 4/1. 2010. pp. 719–729.
8. https://ru.wikipedia.org/the accident at Three Mile Island.
9. Russell B. *A History of Western Philosophy and Its Connection with Political and Social Circumstances from the Earliest Times to the Present Day*. New York: Simon and Schuster. 1945. 834 p.
10. Ledenev V.V., Odnolko V.G. Analysis of the causes of accidents of buildings and structures and ways to increase their reliability. *Bulletin of TSTU*. Vol. 18. No. 2. 2012. pp. 449–457.
11. Vasiliev G.G., Salnikov A.P. Analysis of the causes of accidents of vertical steel tanks. *Oil Industry*. No. 2. 2014. pp. 106–108.
12. Гордон Б.Г. http://nsrus.ru/materialy/stati/gordon-b-g-uroki-avarii-na-atomnyh-stancijah.html.
13. Balatsky E.V. The concept of scaling social phenomena N. Taleb and its application. *Social Sciences and Modernity*. No. 4. 2010. pp. 139–150.

8

Reliable Recommender System Using Improved Collaborative Filtering Technique

Rahul Katarya

Delhi Technological University

CONTENTS

8.1 Introduction: Background and Driving Forces 113
8.2 Related Work 114
8.3 Proposed Work 115
8.4 Experiment and Results 116
8.5 Conclusion 117
References 118

8.1 Introduction: Background and Driving Forces

Recommender system has become a significant part of the e-commerce industries in the recent years. The recommendation system is a subpart of the information filtering system, which aims to facilitate users by presenting them products related to their choices by predicting the "rating" [1–5]. In this chapter, we have presented a movie recommender system, which employs the average ratings of the users to analyze and predict the value of a particular movie. We used well-known datasets as MovieLens, Jester, and Douban to analyze our proposed recommender system. We examined the proposed recommender system with various systems using several evaluation metrics such as root mean square error (RMSE) and mean absolute error (MAE) and found that the proposed system performed better than the already existing system. The use of recommender system has changed the behavior of online customers and websites. It has brought about a more user-friendly experience for the users by presenting them products related to their own choice of selection. It makes predictions based on the user's history of purchases or ratings. Recommendation system has found its applications in various fields such as news, books, research articles, search queries, home shopping, and movie selection [6–11]. Recommender systems make predictions based on one of the two methodologies such as content-based filtering (CB) or collaborative filtering (CF) [12–16]. The basic idea of CF is that it builds a model based on the users' past performance as well as similar decisions made by others. It simply means that if person A and person B have the same opinion in one field, they are most likely to have a similar opinion in the other field as well. The key advantage of CF is that it only considers the huge amount of data available and is independent of the machine analyzable content. This simply means that no extra information is needed about the product whose rating is to be predicted and just the value of one parameter is essential. Therefore, CF has its wide use in the recommendation system. The most commonly used algorithms and evaluation metrics are K-means, cosine, Pearson, Manhattan, and Spearman distances. Following are some of the major contributions:

- The algorithm is a self-devised approach to predict ratings.
- The algorithm uses a two-level filtering process to know about the nearest neighbors.
- The results are better than the earlier existing systems.
- The RMSE rate is 3%–37% better than the other algorithms.
- The MAE rate is 3%–38% better than the other algorithms.

The remainder of the chapter is divided into the following sections: related work is discussed in Section 8.2, suggested works is mentioned in Section 8.3, and experimental outcomes are discussed in Section 8.4. Finally, conclusion and forthcoming work are included in Section 8.5.

8.2 Related Work

Recommender systems play a vital role in human life, helping the humans every day by recommending various items in the online environment. Various e-commerce websites such as Amazon.com, Flipkart.com, Pepperfry.com, and Lenskart.com provide tremendous sources of knowledge, and we choose specific products based on various filtering techniques. In pattern recognition, the k-nearest neighbors algorithm (k-NN) is a non-parametric method used for classification and regression. Two vectors, each of length equal to the number of movies, have been made, one of which stores the total rating of the particular movie and the other stores the number of ratings. Various works have used the matrix factorization approach for recommender systems. An intelligent recommender system was presented in which the author used fuzzy cognitive maps (FCMs) [18]. Trust is another important factor, which plays a significant role in recommendations. An effective recommender system was presented with four important factors such as trustworthiness, expertise, uncertainty, and cost [19]. In another research work, the mathematical regularities were presented with similarity expressions, with data structure and designed linear time algorithm similarities [20]. Various studies have been done in recommender systems in which authors studied the behavior, algorithms, and techniques [21–23]. Kunaver et al. studied the diversification in recommender systems in which they analyzed some interesting topics of recommender system such as feedback collection, data sparsity, cold start, big data problem, and overfitting [24]. Khan et al. presented the concept of cross-domain recommender systems in which they discussed the various aspects of recommender system such as building blocks, user item overlap, group recommender systems, and multi-domain recommender systems [25]. Kim et al. suggested the latest trends and developments in recommender systems in which they reviewed 20 years' research publications [26]. They analyzed the literature with some major components of recommender system such as data mining, machine learning, electronic commerce, algorithms, information retrieval, and social networks.

8.3 Proposed Work

In Figure 8.1, a system based on the P-distance algorithm is applied in the CF environment. In the first step, the total rating of each movie and the total number of ratings for each movie are calculated and stored in a table. If the dataset has m number of users and n number of movies, then this calculation is done in $O(mn)$ time complexity. In the second step, each movie is tested to find out whether it is the neighbor of the movie whose rating is to be predicted or not. This calculation is done by using the cosine distance, Pearson distance, and so on. Now, if the movie is not the neighbor, we check the next movie. Once it has been found that the movie is a neighbor, we move on to the third step. In this step, further filtering is done if the present neighbor is eligible for being a neighbor by checking the deviation of the average rating of the movie with its actual rating. If is not eligible, we scan the next movie in the dataset. If it is still a neighbor after level-2 filtering, it is now given the reputation of an actual neighbor. Since we have found a neighbor, we now compute the predicted value by taking the average of the ratings. In addition, we update the total rating of each movie table and the total number of ratings for each movie table. Further, we calculate the MAE and RMSE to analyze the error between the predicted and the actual value. The proposed technique is very efficient than the conventional approaches for recommendation systems as it reduces the time complexity and predicts the ratings of the movies better than the conventional approaches. Since we have finally

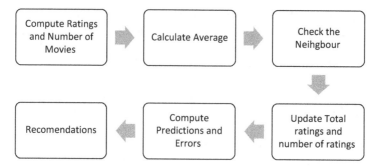

FIGURE 8.1
Workflow of a proposed recommender system.

found the neighbor, the prediction is made as to what might be the rating of the required movie.

8.4 Experiment and Results

In this section, we will discuss the results and experiments, which have been performed with the utilization of three well-known public datasets such as MovieLens, Douban, and Jester. The calculated MAE and RMSE are presented in Figures 8.2 and 8.3.

$$\text{MAE} = \frac{\sum |r_{ui} - r'_{ui}|}{N} \tag{8.1}$$

where
u = user
i = item
r_{ui} = rating of user u on item i
r'_{ui} = predicted rating of user u on item i
N = total number of predictions
RMSE is defined as follows:

$$\text{RMSE} = \sqrt{\frac{\sum_{(i,\alpha)\in E^{Test}} \left(r_{i\alpha} - r_{i\alpha}^{-}\right)^2}{\left|E^{Test}\right|}} \tag{8.2}$$

where
$r_{i\alpha}$ = predicted rating
$r_{i\alpha}^{-}$ = actual rating in the summation

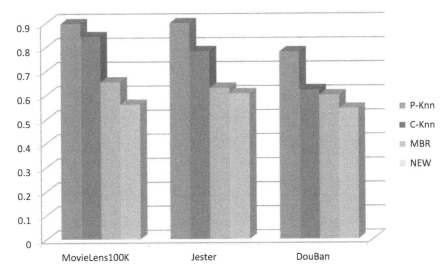

FIGURE 8.2
Comparison of MAE results with the proposed system.

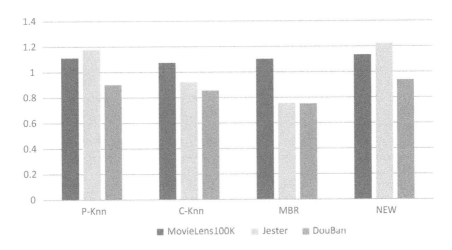

FIGURE 8.3
Comparison of RMSE results with the proposed system.

8.5 Conclusion

The main goal of this chapter is to build a reliable recommendation system, which is better than the other existing approaches in terms of accuracy and time complexity. Recommendation system can be made further accurate by using CB, which takes a number of parameters to recommend the product to the user and not just the rating. We also find that as the sparsity of the dataset increases, the system becomes less reliable as it can be seen from the higher value of MAE for Jester dataset, which is more sparse than MovieLens dataset.

References

1. M. Kunaver and T. Požrl, Diversity in recommender systems—A survey, *Knowledge-Based Syst.*, vol. 123, pp. 154–162, 2017.
2. B. Lika, K. Kolomvatsos, and S. Hadjiefthymiades, Facing the cold start problem in recommender systems, *Expert Syst. Appl.*, vol. 41, no. 4, pp. 2065–2073, 2014.
3. D. Kotkov, S. Wang, and J. Veijalainen, Knowledge-based systems a survey of serendipity in recommender systems, *Knowledge-Based Syst.*, vol. 111, pp. 180–192, 2016.
4. R. Zhang, H. Bao, H. Sun, Y. Wang, and X. Liu, Recommender systems based on ranking performance optimization, *Front. Comput. Sci. China*, vol. 10, pp. 1–11, 2015.
5. M. G. Manzato, M. a. Domingues, A. C. Fortes, C. V. Sundermann, R. M. D'Addio, M. S. Conrado, S. O. Rezende, and M. G. C. Pimentel, Mining unstructured content for recommender systems: an ensemble approach, *Inf. Retr. J.*, vol. 19, no. 4, pp. 378–415, 2016.
6. J. E. Rubio, C. Alcaraz, and J. Lopez, Recommender system for privacy-preserving solutions in smart metering, *Pervasive Mob. Comput.*, vol. 41, pp. 205–218, 2017.
7. R. Katarya and O. P. Verma, Efficient music recommender system using context graph and particle swarm, *Multimedia. Tools Appl.*, vol. 77, no. 2, pp. 2673–2687, 2017.
8. R. Katarya and O. P. Verma, Effectual recommendations using artificial algae algorithm and fuzzy c-mean, *Swarm Evol. Comput.*, vol. 36, pp. 52–61, 2017.
9. R. Katarya and O. P. Verma, Recommender system with grey wolf optimizer and FCM, *Neural Comput. Appl.*, 2016. https://doi.org/10.1007/s00521-016-2817-3.
10. R. Katarya and O. P. Verma, Recent developments in affective recommender systems, *Phys. A Stat. Mech. its Appl.*, vol. 461, pp. 182–90, 2016.
11. V. Bolón-Canedo, N. Sánchez-Maroño, and A. Alonso-Betanzos, Recent advances and emerging challenges of feature selection in the context of big data, *Knowledge-Based Syst.*, vol. 86, pp. 33–45, 2015.
12. Y. Li, C. X. Zhai, and Y. Chen, Exploiting rich user information for one-class collaborative filtering, *Knowl. Inf. Syst.*, vol. 38, no. 2, pp. 277–301, 2014.
13. H. Xia, B. Fang, M. Gao, H. Ma, Y. Tang, and J. Wen, A novel item anomaly detection approach against shilling attacks in collaborative recommendation systems using the dynamic time interval segmentation technique, *Inf. Sci. (Ny).*, vol. 306, pp. 150–165, 2015.

14. D. Sánchez-Moreno, A. B. Gil González, M. D. Muñoz Vicente, V. F. López Batista, and M. N. Moreno García, A collaborative filtering method for music recommendation using playing coefficients for artists and users, *Expert Syst. Appl.*, vol. 66, pp. 234–244, 2016.

15. F. Zhang, T. Gong, V. E. Lee, G. Zhao, C. Rong, and G. Qu, Fast algorithms to evaluate collaborative filtering recommender systems, *Knowledge-Based Syst.*, vol. 96, no. 2016, pp. 96–103, 2016.

16. M. Nasiri and B. Minaei, Increasing prediction accuracy in collaborative filtering with initialized factor matrices, *J. Supercomput.*, vol. 72, no. 6, pp. 2157–2169, 2016.

17. M. E. I. Zheng, F. A. N. Min, H. Zhang, and W. Chen, Fast recommendations with the M-distance, vol. 4, pp. 1464–1468, 2016.

18. J. Aguilar, P. Valdiviezo-dı, and G. Riofrio, A general framework for intelligent recommender systems, *Appl. Comput. Informatics*, vol. 13, pp. 147–160, 2017.

19. W. Jiang, J. I. E. Wu, and G. Wang, On selecting recommenders for trust evaluation, vol. 15, no. 4, 2015.

20. F. Zhang, T. Gong, V. E. Lee, G. Zhao, C. Rong, and G. Qu, Fast algorithms to evaluate collaborative filtering recommender systems, *Knowledge-Based Syst.*, vol. 96, pp. 96–103, 2016.

21. M. J. Cobo, M. a. Martínez, M. Gutiérrez-Salcedo, H. Fujita, and E. Herrera-Viedma, 25years at Knowledge-Based Systems: A bibliometric analysis, *Knowledge-Based Syst.*, vol. 80, pp. 3–13, 2015.

22. J. Bao, Y. Zheng, D. Wilkie, and M. Mokbel, Recommendations in location-based social networks: a survey, *Geoinformatica*, vol. 19, pp. 525–565, 2015.

23. Y. Guan, S. Cai, and M. Shang, Recommendation algorithm based on item quality and user, *Front. Comput. Sci. China*, vol. 8, no. 2, pp. 289–297, 2014.

24. M. Kunaver and T. Požrl, Knowledge-based systems diversity in recommender systems—A survey, vol. 123, pp. 154–162, 2017.

25. M. M. Khan, R. Ibrahim, and F. Computing, Cross domain recommender systems : A systematic, vol. 50, no. 3, pp. 1–34, 2017.

26. M. C. Kim and C. Chen, A scientometric review of emerging trends and new developments in recommendation systems, *Scientometrics*, vol. 104, pp. 239–263, 2015.

9

Categorization of Vulnerabilities in a Software

Navneet Bhatt, Adarsh Anand
University of Delhi

Deepti Aggrawal
Jaypee Institute of Information Technology, Noida

Omar H. Alhazmi
Taibah University

CONTENTS

9.1 Introduction...121
         9.1.1 Vulnerability Types 123
9.2 Literature Review...124
9.3 Modeling Framework..127
         9.3.1 Categorization of Vulnerabilities................................127
         9.3.2 Number of Vulnerabilities in Each Category 130
9.4 Numerical Illustration..132
9.5 Conclusion 134
References...134

9.1 Introduction

Security breaches are of great concern for the well-being of national security and economic growth. The level of security shield of a software system depends on the number of loopholes present in the system and the schedule of patches. In the Department of Homeland Security advisory alert (US–CERT, 2015), the emergency readiness team reported that systems deployed with unpatched versions of Adobe, Microsoft, Oracle, or OpenSSL software are regularly breached. Many of the attacks are due to the fact that malicious users try to exploit the weaknesses present in the system. The Symantec Internet Security Threat Report stated an increase of 23% in the total number of identities breached in the year 2015 and reported a significant growth of 125% in the zero-day vulnerabilities cases. The report further mentioned that in year 2015, there were over one million web attacks each and every day. It is believed that accessing an authentic website will keep the users safe from the cybercriminals. But many times the website administrator fails to secure the system, allowing the exploiters to take advantage of vulnerabilities present in the system to breach the users. Recent trends have shown that about 75% of all authentic website systems have unpatched software vulnerabilities (Symantec, 2016). Therefore, assessing the threat level due to any breach associated with security vulnerabilities is required to address more aggressively.

Generally, a vulnerability appears to enter distinct states during its lifetime. Its life cycle describes a fixed and linear advancement from one stage to the subsequent stages. Like any other product, software vulnerability goes through transition from discovery to patching a fix on an infected system. Specifically, it consists of states: birth, discovery, disclosure, patch release, publication, and automation of the exploitation. Intuitively, a life cycle curve generally follows a bell-shaped curve, and the vulnerability curve resembles the impact of intrusions into a computer system over the time. With the discovery of a vulnerability, the degree of intrusions increases with the rate at which the news of the vulnerability spreads to malicious users. The intrusion rate rises until the developer releases a patch or workaround, and the trend declines after it. The vulnerability discovery process under consideration is deployed after the release of a software. When someone discovers security implications in a software system, the flaw becomes a vulnerability. The discovery is irrespective of the malicious or benign user's intention.

The discovery process attempts to classify all the potential vulnerabilities not at the same time. The vulnerabilities detected over time can be categorized based on the common vulnerability scoring system (CVSS) that is deployed to calculate the severity rating of each vulnerability. The scoring procedure is governed by the Forum of Incident Response and Security Teams and aims to remediate those vulnerabilities that pose the highest risk by providing a normalized score across all hardware and software platforms. The CVSS classifies the vulnerabilities based on various characteristics to assess the severity of a vulnerability. The scoring system assists in investigating several parameters that quantify the intrinsic and penetration capabilities of a vulnerability for breaching a loophole. The CVSS utilizes both qualitative and quantitative characteristics to measure the impact of vulnerability in a software product. The National Vulnerability Database organized by CVE Details (www.cvedetails.com) provides the score report and type of each vulnerability present in a software product. There are about ten types of vulnerabilities that are majorly discovered and are attributed by CVE Details. A brief review of various vulnerability types is given in Section 9.1.1.

9.1.1 Vulnerability Types

The vulnerabilities have been classified based on their underlying behavior during the exploitation process. The CVE Details repository recognized more than 85,500 vulnerabilities from 1999 to 2017. The various types of vulnerabilities are described as follows:

Denial of Service (DoS): It is a cyberattack that occurs when an attacker takes action that makes a machine or network resource temporarily or indefinitely unavailable from accessing the computer system.

Code Execution: It occurs when an attacker remotely executes any command by exploiting a flaw in the system. The attacker introduces a malware and exploits the vulnerability on the target machine remotely.

Memory Overflow: It is a type of leak of unintentional form of memory consumption that occurs when a computer program incorrectly allocates memory and fails to release an allocated block of memory which is no longer needed.

Memory Corruption: It can be described as the vulnerability that may occur when the contents of a memory location are unintentionally altered without an explicit assignment.

Structured Query language (SQL) Injection: It refers to an injection technique, used by an attacker to execute malicious SQL statements that are inserted into an application's database server.

Cross-Site Scripting (XSS): It is a type of injection, in which an attacker can execute malicious scripts into a legitimate website or web application.

Directory Traversal: It is a Hypertext Transfer Protocol (HTTP) exploit that allows attackers to access files and directories and execute commands outside the web root folder.

HTTP Response Splitting: It is a type of vulnerability that allows an attacker to input a single HTTP request to a web server to generate an output stream, and the output stream is inferred as two responses instead of one. This allows the attacker to access the web application.

Gain Information: This type of vulnerability allows a local or remote attacker to gain privileges via a malicious program in the affected application.

Gain Privilege: It is a type of vulnerability that grants the attacker elevated access to the network and its associated data and applications.

Cross-Site Request Forgery: It is an attack that advocates an end user to execute unwanted actions on a web application in which they're currently authenticated.

File Inclusion: It is a type of vulnerability that allows an attacker to access unauthorized or sensitive files available on the web server or to execute malicious files on the web server.

These abovementioned vulnerabilities have different proportion in a software based on the type of software system. A software product based on operating system consists of certain types of vulnerabilities that have higher likelihood of detection as compared to a software of web applications type. As a whole, in the CVE Detail repository, about 30% of vulnerabilities are of code execution type and 22% correspond to the DoS type. The other types such as Overflow, Memory Corruption, and SQL Injection. Captures the leftover proportion. The impact of these vulnerability types assists the discovery process at various time points. From the past vulnerability disclosers, it has been observed that certain type of vulnerabilities share a huge proportion in the discovery process. The proportion of vulnerabilities discovered with respect to time or various types may categorize the discovery process and consequentially helps the developer in identifying the vulnerabilities based on their behavior in each category. This categorization may help the developer for the development of various attributes like: deployment of patch, scheduling active resources in the discovery, and determination of patch release time.

In this chapter, we list out the related research that has been done in the field of vulnerability discovery modeling in Section 9.2. In Section 9.3, we describe how the vulnerabilities of similar nature irrespective of the types are discovered and are consolidated in a time interval according to the noncumulative vulnerability discovery at any instant and present a mathematical derivation for the determination of different vulnerability categories based on time. Next, Section 9.4 provides a numerical background to the proposed categorization of vulnerabilities using different data sets. In Section 9.5, we conclude this work followed by the references.

9.2 Literature Review

The classification of vulnerability discovery models (VDMs) is grouped in time-based and effort-based modeling. The prior captures the vulnerabilities discovered with respect to time, and the latter predicts the vulnerabilities based on the efforts applied. The criteria to predict the vulnerabilities considered time as the governing factor, which was also a major attribute of most of the VDM papers in the literature. Of late, the popularity of VDMs has begun to rise among researchers, but Anderson's thermodynamic model was the one that first familiarized the significance of VDM to capture the vulnerability trend. The outline of the Anderson (2002) model is explicitly based on the Software Reliability Growth Model (SRGM) concept. However, the empirical results stated that the data fitted the worst. But some authors argued that the variation in the fitting is due to a significant drop in the discovery rate and can be attributed to the losing attractiveness over time. Later, Rescorla (2005) classified the vulnerability trend over time following a linear pattern and an exponential pattern. The models were fitted to the data for Red Hat Enterprise Linux 7. However, he was unable to test the predictive accuracy against the Red Hat data and majority of operating systems: WinNT4, Solaris 2.5.1, and FreeBSD 4.0 vulnerability data fit neither a linear nor an exponential model. Furthermore, Alhazmi and Malaiya (2005) proposed an s-shaped logistic model (Alhazmi Malaya Model (AML)) that discussed the behavior of the vulnerability discovery process at different stages of time. At first, the target system is studied in order to infiltrate, and consequently, few vulnerabilities are discovered. Over a period of time, users understand the system and begin targeting the software, resulting in an amplified discovery of vulnerabilities. Finally, the discovery process flattens due to the availability of newer technology that encourages the users in switching to newer versions and the disinterest in finding the loopholes.

Alhazmi and Malaiya (2005) also considered the impact of efforts during the discovery process. They considered changes due to environmental factors such as number of installations, which directly relate the need of more efforts to exploit vulnerabilities when the share of installed base is in larger number. Furthermore, Alhazmi et al. (2007) deduced a linear vulnerability discovery (LVD) model as an approximation of AML model to capture the trend in vulnerabilities, more specifically of newer software versions. Moreover, they demonstrate some useful information regarding the maximum vulnerability discovery rate by mathematically analyzing the model. Later, Joh et al. (2008) advocated asymmetric behavior of the discovery curve and used Weibull distribution for the discovery prediction. Similarly, Younis et al. (2011) modeled the discovery process using the folded normal distribution. Joh and Malaiya (2014) modeled the skewness present in the vulnerability discovery rate by introducing some s-shaped models using the Weibull, beta, gamma, and normal distributions. Lately, Kapur et al. (2015) accounted to use logistic detection rate to predict the potential vulnerabilities present in a software. Recently, Anand and Bhatt (2016) discussed a special behavior of discovery rate that follows a hump-shaped curve, and that causes the vulnerability exposure according to the attractiveness of the software adoption in the market. Joh and Malaiya (2016) considered the periodicity aspect to assess the vulnerability at a time. They examined the behavior of different software products, and specifically, the Microsoft products encountered an increase in the vulnerability discovery rate during the mid-year periods. Some authors have considered the multi-version aspect of the vulnerability discovery phenomenon. Kim et al. (2007) considered the impact of shared code among the successive versions and proposed a modeling framework for the vulnerability discovery in multi-version software. Recently, Anand et al. (2017) suggested a methodology to anticipate the vulnerability exposures in the case of multi-version software system.

A study done by Ozment (2007) analyzed the detection process of vulnerability as a dependent process. He mentioned three major causes for dependency in the vulnerabilities discovery process: a new type of vulnerability discovery, a new target location that was previously unconsidered, or a new tool for detecting vulnerabilities. Moreover, the dependency of vulnerabilities is associated with a time to next vulnerability (TTNV) and indeed the reason conforming a vulnerability to be dependent is a small TTNV as they are detected soon after the discovery of another vulnerability. Later, another study proposed by Bhatt et al. (2017) inculcated the phenomenon of the additional vulnerability discovery by comprising the impact of previously discovered vulnerabilities. They separated the discovery of vulnerabilities in two different forms: normal discovery and additional discovery. During the vulnerability discovery process, a certain proportion of vulnerabilities are discovered due to the influence of previously discovered vulnerabilities. Such vulnerabilities are termed as additional vulnerabilities and the previously discovered vulnerabilities correspond to leading vulnerabilities. The behavior of the cumulative vulnerability discovery function comprising this behavior follows an s-shaped curve. The noncumulative discovery function or the vulnerability discovery rate upon investigation resulted in a bell-shaped curve representing the behavior of vulnerabilities discovery under various time states. The initial time corresponds to a quick increase in the discovery following an increase in the detection rate. This increase reaches its peak and then substantially decreases over the software life time. They mathematically analyzed the vulnerability discovery rate and deduced some important information. The noncumulative vulnerability distribution function attains its peak, that is, the time point at which the vulnerability discovery rate is maximum can be determined. Moreover, the proportion of vulnerabilities that are previously discovered and additional discovered can be separately determined.

This article attempts to categorize the vulnerability count of a software product being discovered after the release of a software. The modeling framework discussed here has extended the ideology proposed by Bhatt et al. (2017) to categorize the vulnerabilities encountered throughout its life cycle. When a software is readily available for its release, prior expectation of its announcement fetches out more numbers of testers—either benign or malicious—that are eagerly waiting to exploit the loopholes due to its increased attractiveness. Due to this rising popularity, a significant amount of vulnerabilities are detected and consequently increase the vulnerability detection rate. The vulnerability discovery rate attains a peak after which the rate declines, suggesting the discovery of vulnerabilities at a much slower rate.

The framework proposed here to classify vulnerabilities has its roots embedded in software reliability engineering. The present methodical approach is based on the work proposed by Kapur et al. (2000) in the context of categorizing errors in a software and by Mahajan et al. (1990) in the context of adopter categories. With their analogy we have classified vulnerabilities over the life cycle of a software based on the time they take during the discovery process. It is well known that not all vulnerabilities are detected instantaneously and the discovery of one vulnerability triggers the detection of additional vulnerabilities. The discovery truly depends on the efficiency and effectiveness of a tester trying to capture the loopholes. These vulnerabilities get discovered based on the time elapsed in the discovery process. The vulnerabilities following each category would be based on the likelihood of the discovery. This study will help the tester to forecast the vulnerabilities present in a software system and allocate the efforts based on the number of vulnerabilities discovered during each category.

9.3 Modeling Framework

The VDM proposed by Kapur et al. (2015) inculcates the phenomenon of involving logistic rate to formulate the discovery process. In line with them, Bhatt et al. (2017) proposed a modeling framework that was governed by two factors: the first factor represents the vulnerability discovery with a rate r and the second factor corresponds to the discovery of additional vulnerabilities with a rate s. The vulnerability discovery process as discussed by Kapur et al. (2015) and Bhatt et al. (2017) can be described as the following differential equation:

$$\frac{d\Omega(t)}{dt} = r\left(N - \Omega(t)\right) + s\frac{\Omega(t)}{N}\left(N - \Omega(t)\right) \tag{9.1}$$

where r is the rate of vulnerability detection, s represent the rate that triggers the discovery of additional vulnerabilities due to the previously disclosed vulnerabilities, N represents the potential number of vulnerabilities present in the software, $\Omega(t)$ represents the cumulative vulnerabilities discovered by time t, the count $(N-\Omega(t))$ denotes the undetected vulnerabilities lying dormant in the software, and $\frac{\Omega(t)}{N}$ is the fraction of discovered vulnerabilities by time t. By solving the differential equation (9.1) using the initial condition $\Omega(t = 0) = 0$, a closed form solution can be obtained as

$$\Omega(t) = N\left(\frac{1 - e^{-(r+s)t}}{1 + \frac{s}{r}e^{-(r+s)t}}\right) \tag{9.2}$$

The cumulative number of vulnerabilities given by Equation (9.2) in a software system will follow an s-shaped growth curve.

9.3.1 Categorization of Vulnerabilities

The proposal by Bhatt et al. (2017) required some further analysis and by mathematically analyzing the model we can deduce some important useful characteristics from their proposal. For example, the maximum vulnerability discovery rate can be determined analytically and can be helpful in understanding the nature of vulnerabilities present in the software. Hence, the noncumulative instantaneous vulnerability discovery rate at any instant t is given by the first derivative of Equation (9.2), wherein the rate of vulnerability discovery is

$$\Omega'(t) = \frac{d}{dt}\Omega(t) = N\frac{r(r+s)^2 e^{-(r+s)t}}{[r + se^{-(r+s)t}]^2}$$ (9.3)

The curve of the noncumulative vulnerability discovery rate, $\Omega'(t)$, is depicted in Figure 9.1. The highest vulnerability discovery rate occurs at the midpoint at

$$T^* = -\frac{1}{(r+s)}\ln(r/s)$$ (9.4)

If $r=s$, then $T^*=0$ and if $r>s$, then T^* doesn't exist. Hence, based on the values of r and s, T^* can take positive, zero, or negative values. The negative value of T^* implies that the noncumulative curve is decreasing in nature.

The vulnerability discovery rate at the peak value can be obtained by plugging the value of T^* from Equation (9.4) in Equation (9.3).

$$\Omega'(T^*) = \frac{N}{4s}(r+s)^2$$ (9.5)

Figure 9.1 of noncumulative vulnerability discovery curve is symmetric with respect to time. It can be validated by checking $\Omega'(t=0) = \Omega'(t=2T^*) = N \cdot r$, that is, the discovery of noncumulative vulnerabilities is symmetric around the peak time T^* up to $2T^*$ and the discovery process is initiated by a specific count of vulnerabilities at the time of release of the software. This is because some vulnerabilities are reported before the release of a software, due to the existence of shared code or due to mismatch with the system configuration.

Following the given observations of the model, we further examine the vulnerability discovery process to categorize the vulnerabilities by inspecting the trend in both instantaneous vulnerability discovery rates $\Omega'(t)$ and its rate of change $\frac{d\Omega'(t)}{dt} = \Omega''(t)$. These trends indicate the nature of

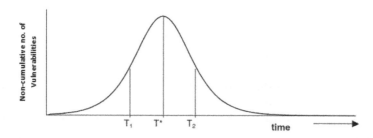

FIGURE 9.1
Noncumulative vulnerability curve.

TABLE 9.1

Trend in the Instantaneous Discovery Rate

Time Interval	Trend in $\Omega'(t)$
Zero to T_1	Increasing at an increasing rate
T_2 to T^*	Increasing at a decreasing rate
T^* to T_2	Decreasing at an increasing rate
Beyond T_2	Decreasing at a decreasing rate

vulnerabilities upon its discovery with respect to the time of detection. The trends for the VDM given in Equation (9.2) are summarized in Table 9.1 as follows:

Vulnerabilities can be categorized based on various metrics such as severity based on the impact, vulnerability types, or relative easiness during the discovery process. In our study, the latter case is exclusively considered for the classification. Vulnerabilities are categorized according to the time they take in their discovery. It is a common practice that during the discovery process, vulnerabilities are searched in paths or locations that are more executed in the operational environment rather than in paths traversed under uncommon or infrequent conditions. When a vulnerability is discovered while investigating the codes for the identification of the causes, it may result in the discovery of other vulnerabilities. The number of vulnerabilities found consequent to a loophole gives an idea about the type of vulnerabilities being discovered and the nature of remaining vulnerabilities. If the given information is available to the software developer, required efforts can be streamlined to attain a better discovery rate and suitable control strategies or patch management can be initiated for a particular vulnerability category.

In real-life situations, it has been observed that a large number of vulnerabilities are easily detected during the initial stage of discovery process. Due to the change in usage environment, and specifically the software, systems with shared code will experience a swift growth in vulnerabilities with respect to its release and will attract the attention of users either benign or malicious to break the system. In this case, the vulnerability discovery rate at the beginning has a high value as compared to the value at the end of the software life cycle due to saturation. But according to the trend observed in the instantaneous discovery rate, the discovery rate increases in the interval $(0,T^*)$, with an increasing rate in $(0,T_1)$ interval. The increase in the discovery rate may be attributed to the fact that the discovery process increases the ability of testers, claiming an increase in their efficiency. The increase also gives rise to the discovery of additional vulnerabilities. After reaching time T_1, the discovery process still increases but with a decreasing rate. The vulnerabilities discovered during the interval (T_1,T^*) have a significant impact from the previously discovered vulnerabilities. During this interval, the software product is adopted by a large number of users, and consequently the vulnerabilities reported from external sources will impact the discovery process. The vulnerabilities detected during the interval $(0,T^*)$ are relatively easier to locate and those discovered later as comparatively difficult vulnerabilities. The point T^* is the inflection point for the cumulative vulnerability curve $\Omega(t)$ and hence it resembles as the point of maxima for the noncumulative vulnerability discovery rate, $\Omega'(t)$. In the interval (T^*,T_2), the vulnerability discovery rate decreases, that is, lesser number of vulnerabilities are discovered due to the disinterest shown by users when the system starts to be replaced by newer versions. The drop in the discovery rate may also be attributed to the availability of competitive software products with more features. The firm during this time internal may also release an upgraded version of the software. And the vulnerabilities discovered may be attributed to the code shared with the next version. In the interval (T_2,∞), the vulnerability discovery rate decreases with decreasing rate and the noncumulative curve become asymptotic to the time axis. The vulnerabilities belonging to this category are very hard to locate, and moreover, people also show a disinterest in finding them; hence, the noncumulative curve flattens at the end.

9.3.2 Number of Vulnerabilities in Each Category

To implement the proposed methodology for developing the vulnerability categories, we find the size of each category. The point of inflection or the transition points T_1 and T_2 can be found by taking a derivative of $\Omega'(t)$. When the second derivative of $\Omega(t)$ is equated to zero, the transition points occurs at times t equal to

$$T_1 = -\frac{1}{(r+s)}\ln\left[(2+\sqrt{3})\frac{r}{s}\right]$$

(9.6)

$$T_2 = -\frac{1}{(r+s)}\ln\left[\left(\frac{1}{(2+\sqrt{3})}\right)\frac{r}{s}\right]$$

(9.7)

Using T_1 and T_2, the number of vulnerabilities of each category present in the software can be computed.

The first category consists of those number of vulnerabilities that are discovered when the discovery rate suggests an increasing rate and can be given as

$$V_1 = \Omega(T_1) = N\left[\frac{1}{3+\sqrt{3}} - \frac{2+\sqrt{3}}{3+\sqrt{3}}\cdot\frac{r}{s}\right] - r\cdot N$$

(9.8)

These number of vulnerabilities are discovered after a software is released, and due to the attractiveness of the software, the vulnerability discovery rate constitutes an increasing fashion and hence discovers a large proportion of potential vulnerabilities. Moreover, the vulnerabilities discovered will have a large proportion of certain types of vulnerabilities having a high likelihood of discovery of those easily detectable.

The vulnerabilities identified during the second category will have a majority of vulnerabilities located due to the impact of installed base. The vulnerabilities discovered have a significant proportion of users' reported flaws. The number of vulnerabilities that belong to this category can be computed as

$$V_2 = \Omega(T^*) - \Omega(T_1) = \frac{N}{2\sqrt{3}}\left[1 + \frac{r}{s}\right] \tag{9.9}$$

After the software reaches the peak of its popularity, at time point T^*, the vulnerability discovery rate starts to decrease. The vulnerabilities are reported or discovered in a decreasing fashion, and by the time it reaches the time point T_2, the saturation phase starts. An equal number of vulnerabilities are likely to be discovered during the interval (T_1, T^*) due to the symmetry of the noncumulative curve. The vulnerabilities constituted during this time interval are mainly due to availability of the new version that shares its code with the previous version. The count of vulnerabilities captured in the interval (T^*, T_2) can be given as

$$V_3 = \Omega(T_2) - \Omega(T^*) = \frac{N}{2\sqrt{3}}\left[1 + \frac{r}{s}\right] \tag{9.10}$$

Finally, the vulnerability discovery rate slows down, or the cumulative number of vulnerabilities asymptotically approaches its highest value. These group of vulnerabilities are identified during the last phase of the software life cycle. The vulnerabilities discovered during this period are significantly less and constitute after-life vulnerabilities because the software reaches its end life. Moreover, the vulnerabilities discovered during this period would be affecting a certain number of users that have not been switched to the newer version and the installed base of the software would be very low. The vulnerabilities count that lies beyond time point T_2 is given as

$$V_4 = N - \Omega(T_2) = \frac{N}{3 + \sqrt{3}}\left[1 + \frac{r}{s}\right] \tag{9.11}$$

The vulnerability categorization is based on the points of inflection of the VDM, which depends on the values of r and s. Hence, for some growth curve, there may be a possibility that one or more of T_1, T^*, and T_2 may or may not exist, and accordingly, the vulnerability categories may change.

9.4 Numerical Illustration

This section deals with the empirical results deduced using the proposed methodology for the vulnerability categorization. Here, we evaluate the VDM proposed by Bhatt et al. (2017) and then calculate the desired inflection points for the determination of number of vulnerabilities in each category. For the estimation procedure, nonlinear least square method is used to evaluate the parameters of the model. The methodology is illustrated by considering four different data sets belonging to different operating systems of two product families, namely, Apple Macintosh and Microsoft Windows (Mac Os X, 2016; Mac Os X Server, 2016; Windows XP, 2016; Windows Server, 2016). Table 9.2 summarizes the parameter estimates for all the data sets considered.

The noncumulative curves are presented in Figure 9.2 for all the four vulnerability data sets. And the inflection points of the noncumulative curves are calculated and given in Table 9.3. It is interesting to note that the noncumulative curves for the vulnerability process are symmetric in nature due to the fact that the vulnerability discovery function under consideration follows an S-shaped curve and its respective noncumulative vulnerability intensity curve is bell shaped. The inflection points given in Table 9.3 indicate the time at which the behavior of vulnerability discovery intensity function changes its performance. After each time point, the likelihood of vulnerability discovery changes. For Windows XP, the time interval corresponding to $(0, T_1)$ impact the discovery process with an increasing rate and constitute those vulnerabilities that have a higher likelihood of discovery. In the time interval (T_1, T^*), due to the learning phenomenon in the S-shaped curve, the vendor/reporters understand the system and its market share also increases, implying a rapid discovery of many vulnerabilities. After attaining its peak at T^*, the discovery rate starts to decrease and due to the symmetricity of the curve, an equal number of vulnerabilities are discovered that can be viewed in Table 9.4 for the categories V_2 and V_3. Finally, after time point T_2, the vulnerability discovery rate decreases due to the disinterest shown by the users in the discovery. The vulnerabilities that are still present in the system and are not discovered even after the software support ended accounts in the

TABLE 9.2

Model Parameter Estimation Results

Parameter Estimates	Data Set			
	Windows XP	Windows Server 2008	Apple Mac OS	Apple Mac Server
N	942.15	566.13	1599.72	654.85
R	0.0107	0.0217	0.0118	0.0134
s	0.3059	0.6612	0.2338	0.5254

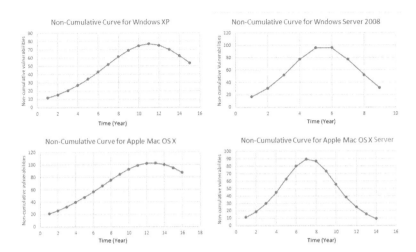

FIGURE 9.2
Noncumulative curves.

TABLE 9.3

Inflection Points

Time Points	Data Set			
	Windows XP	Windows Server 2008	Apple Mac OS X	Apple Mac OS X Server
T_1	6.409678	3.074056	6.788837	4.353977
T^*	10.56833	5.002266	12.14913	6.797624
T_2	14.72698	6.930477	17.50942	9.241271

TABLE 9.4

Number of Vulnerabilities in Each Category

Category	Data Set			
	Windows XP	Windows Server 2008	Apple Mac OS X	Apple Mac OS X Server
V_1	173	105	274	125
V_2	281	168	485	194
V_3	281	168	485	194
V_4	206	397	355	142

category V_4. As in case of Windows XP, the support ended in 2014, but some vulnerabilities are still present that were later discovered in the year 2017.

The number of vulnerabilities in each category are calculated and given in Table 9.4.

9.5 Conclusion

In this chapter, we have performed a mathematical analysis to consolidate different types of vulnerabilities into four categories. The result of this analysis will help the software tester to capture the trend in the discovery rate, which further allows the tester to locate all the paths traversed for the possibility of a vulnerability discovery based on the underlying category. The tester can predict the amount of vulnerabilities that can be discovered during the course of time and knowing the fraction of vulnerabilities present in each category will help him to schedule the resources for identification. Hence, the vendor may know the vulnerability counts that have a high likelihood of discovery or those easily detectable would be targeted prior and may require a less amount of effort in the identification. The mathematical derivation performed in the analysis has been validated in four different operating system data sets.

References

Alhazmi, O. H., & Malaiya, Y. K. (2005). Modeling the vulnerability discovery process. In *16th IEEE International Symposium on Software Reliability Engineering (ISSRE'05)* (10 pp). IEEE.

Alhazmi, O. H., Malaiya, Y. K., & Ray, I. (2007). Measuring, analyzing and predicting security vulnerabilities in software systems. *Computers & Security*, 26(3), 219–228.

Anand, A., & Bhatt, N. (2016). Vulnerability discovery modeling and weighted criteria based ranking. *Journal of the Indian Society for Probability and Statistics*, 17(1), 1–10.

Anand, A., Das, S., Aggrawal, D., & Klochkov, Y. (2017). Vulnerability discovery modelling for software with multi-versions. In Ram, M. and Davim, J. P. (Eds.) *Advances in Reliability and System Engineering* (pp. 255–265). Cham: Springer International Publishing.

Anderson, R. (2002). Security in open versus closed systems—the dance of Boltzmann, Coase and Moore. Technical report, Cambridge University, England.

Bhatt, N., Anand, A., Yadavalli, V. S. S., & Kumar, V. (2017) 'Modeling and characterizing software vulnerabilities', international journal of mathematical. *Engineering and Management Sciences*, 2(4), pp. 288–299.

Joh, H., Kim, J., & Malaiya, Y. K. (2008). Vulnerability discovery modeling using Weibull distribution. In *2008 19th International Symposium on Software Reliability Engineering (ISSRE)* (pp. 299–300). IEEE.

Joh, H., & Malaiya, Y. K. (2014). Modeling skewness in vulnerability discovery. *Quality and Reliability Engineering International*, 30(8), 1445–1459.

Joh, H., & Malaiya, Y. K. (2016). Periodicity in software vulnerability discovery, patching and exploitation. *International Journal of Information Security*, 16(6), 1–18.

Kapur, P. K., Bardhan, A. K., & Kumar, S. (2000). On categorization of errors in a software. *International Journal of Management and System*, 16(1), 37–38.

Kapur, P. K., Sachdeva, N., & Khatri, S. K. (2015). Vulnerability discovery modeling. In *International Conference on Quality, Reliability, Infocom Technology and Industrial Technology Management* (pp. 34–54).

Kim, J., Malaiya, Y. K., & Ray, I. (2007). Vulnerability discovery in multi-version software systems. In *High Assurance Systems Engineering Symposium*, 2007. HASE'07. 10th IEEE (pp. 141–148). IEEE.

Mac Os X. (2016). Vulnerability statistics. http://www.cvedetails.com/product/156/Apple-Mac-Os-X.html?vendor_id=49. Accessed 6 February 2016.

Mac Os X Server. (2016). Vulnerability statistics. http://www.cvedetails.com/product/2274/Apple-Mac-Os-X-Server.html?vendor_id=49. Accessed 6 February 2016.

Mahajan, V., Muller, E., & Srivastava, R. K. (1990). Determination of adopter categories by using innovation diffusion models. *Journal of Marketing Research*, 27(1), 37–50.

Ozment, J. A. (2007). Vulnerability discovery & software security (Doctoral dissertation, University of Cambridge).

Rescorla, E. (2005). Is finding security holes a good idea?. *IEEE Security & Privacy*, 3(1), 14–19.

Symantec. (2016). Symantec internet security threat report. http://www.symantec.com.

US-CERT (2015). Top 30 targeted high risk vulnerabilities. US-CERT Alert (TA15–119A). Available at https://www.us-cert.gov/ncas/alerts/TA15-119A.

Windows Xp (2016). Vulnerability statistics. http://www.cvedetails.com/product/739/Microsoft-Windows-Xp.html?vendor_id=26. Accessed 6 February 2016.

Windows Server 2008. (2016). Vulnerability statistics. https://www.cvedetails.com/product/11366/Microsoft-Windows-Server-2008.html?vendor_id=26. Accessed 20 February 2016.

Younis, A., Joh, H., & Malaiya, Y. (2011). Modeling learningless vulnerability discovery using a folded distribution. In *Proceedings of SAM* (Vol. 11, pp. 617–623).

10

Reliable Predictions of Peak Electricity Demand and Reliability of Power System Management

Caston Sigauke
University of Venda

Santosh Kumar
University of Melbourne

Norman Maswanganyi
University of Limpopo

Edmore Ranganai
University of South Africa

CONTENTS

10.1 Introduction 138
10.2 Methodology 139
    10.2.1 Generalized Additive and Quantile Regression Models 139
    10.2.2 The Proposed AQR Model 141
    10.2.3 Parameter Estimation 142
    10.2.4 Modeling Residual Autocorrelation 143
    10.2.5 Forecasting Electricity Demand When Covariates Are
           Known in Advance 143
    10.2.6 Forecasting Electricity Demand When Covariates Are
           Not Known in Advance (Operational Forecasts) 144
    10.2.7 Forecast Combinations and Error Measures 144
10.3 Empirical Results 145
    10.3.1 Exploratory Data Analysis 145
    10.3.2 Results and Discussions 145
    10.3.3 Results: Operational Forecasts 148
10.4 Point Process Characterization of the Occurrence of Extreme
    Peak Electricity Demand 150
    10.4.1 Nonlinear Detrending: Modeling Extreme Peak Electricity
           Demand (Operational January 1, 2012, to December 31, 2016) .. 150
    10.4.2 Estimation of Threshold 150

10.5 Reliability Analysis of Power Systems..154
    10.5.1 Reserve Margin..154
    10.5.2 Loss of Load Expectation...156
    10.5.3 Exponential Reliability Function................................156
    10.5.4 Discrete-Time Markov Chain Analysis 157
10.6 Conclusion 157
Acknowledgments...158
References..158

10.1 Introduction

Electricity demand forecasting is very important to system operators who plan for hour-to-hour electricity production to meet demand, at the same time ensuring grid stability, and to strategic managers who plan for medium-term risk assessment and capacity expansion. The chapter presents a point process characterization of peak electricity demand. An analysis of power systems reliability in the presence of random peak loads is also discussed in the chapter. Modeling the upper tail distribution of peak electricity demand is important as there are times when high load values exceed the maximum generated electricity, which could cause problems to the system, including grid instability. A severe stress on the grid system occurs when the peak electricity demands' amplitude exceeds a predetermined sufficiently high threshold that is determined using extreme value theory methods (Chidodo and Lauria, 2012). This high load tests the reliability of the electrical power systems components.

Power system reliability is defined as "the probability that an electric power system can perform a required function under given conditions for a given time interval" (Čepin, 2011). Some of the important measures in reliability modeling include reliability indexes such as the loss of load expectation (LOLE), which is derived using daily peak load (Čepin, 2011), including among others the extreme peak load frequency (EPLF), which is the average number of peak loads above a sufficiently high threshold (Chidodo and Lauria, 2012). Chidodo and Lauria (2012) discuss the probabilistic characterization of peak loads. It is assumed that peak load follows a Poisson process. The authors argue that it is essential for system operators to have information about the occurrences of extreme peak loads as these form part of the power components reliability modeling.

Prediction of probabilistic forecasts of future peak electricity demand is discussed in McSharry et al. (2005). The authors argue that the risk management and assessment of uncertainty in the forecasts can be improved. Several methods are discussed in the literature on short-term to long-term electricity demand forecasting. For recent reviews on these techniques, see Hong and Fan (2016) and Hong et al. (2016), among others. Methodologies discussed in the chapter include among others, variable selection, hierarchical forecasting, and weather station selection.

This chapter focuses on long-term peak electricity demand forecasting using additive quantile regression (AQR) models. These models are discussed in the literature. An application of generalized AQR models to electricity demand forecasting is presented in Gaillard et al. (2016). The developed hybrid models ranked first on two tracks, that is, load and price forecasting in the global energy forecasting competition of 2014. The modeling framework proposed by Gaillard et al. (2016) is extended by Fasiolo et al. (2017) who used the Bayesian estimation. To implement the developed models, Fasiolo et al. (2017) developed an R statistical package "qgam". In both chapters, variable selection techniques are not discussed. In this chapter, shrinkage methods for variable selection, and in particular, lasso variable selection via hierarchical interactions are used. On an application of AQR, Maswanganyi et al. (2017) used partially linear additive quantile regression (PLAQR) models in forecasting South African peak electricity demand. It is noted from this study that the model with pairwise hierarchical interactions produces more accurate forecasts. Estimation of the parameters of the developed PLAQR models was done using the maximum likelihood method. If not managed properly the stability of the grid can be disrupted by extreme peak electricity demand (Sigauke and Bere, 2017). Using South African data, the authors present an application of extreme value theory distributions in modeling peak electricity demand.

The chapter discusses an analysis of power systems reliability in the presence of random extreme peak loads. Modeling the upper tail of the distribution of peak loads is important in reliability modeling. There are times when peak electricity demand exceeds the maximum generated electricity. A severe stress on the grid system occurs when the peak electricity demand's amplitude exceeds a predetermined sufficiently high threshold, which is determined using extreme value theory methods. The methodology used in the chapter is discussed in Section 10.2, including variable selection, estimation of parameters, and error measures. The empirical results are presented in Section 10.3, while the point process characterization of extreme peak electricity demand is discussed in Section 10.4. A brief discussion on reliability analysis is given in Section 10.5, and conclusion is provided in Section 10.6.

10.2 Methodology

10.2.1 Generalized Additive and Quantile Regression Models

Generalized additive models (GAMs) developed by Hastie and Tibshirani (1986, 1990) allow flexibility in modeling linear predictors as a sum of smooth functions. Let y_t denote daily peak electricity demand where $t = 1, \ldots, n$ and the corresponding p covariates, $x_{t1}, x_{t2}, \ldots, x_{tp}$. The AQR model discussed in the works of Gaillard et al. (2016) and Fasiolo et al. (2017) is given as follows:

$$y_{t,\tau} = \sum_{j=1}^{p} s_{j,\tau}\left(x_{tj}\right) + \varepsilon_{t,\tau}, \tau \in (0,1) \tag{10.1}$$

where $s_{j,\tau}$ are smooth functions with a dimension denoted by p and $\varepsilon_{t,\tau}$ is the error term. The smooth function, s, is written as

$$s_j(\mathbf{x}) = \sum_{k=1}^{q} \beta_{kj} b_{kj}\left(x_{tj}\right) \tag{10.2}$$

where β_{kj} denotes the unknown parameters and $b_{kj}\left(x_{tj}\right)$ represents the spline base functions with the dimension of the basis being denoted by q. There are several smoothing spline bases ranging from P-splines, thin-plate regression splines, B-splines, cubic regression (CR) splines, to cyclic cubic regression (CCR) splines, among others (Wood, 2017). In this chapter, CR, CCR, and P-splines are used. With CR splines, parameters are interpretable and computationally cheap (Wood, 2017). However, they are knot based and can smoothen with respect to one covariate (Wood, 2017). CCR splines are an extension of CR splines and perform well with data that are cyclic or exhibit seasonality. On the other hand, P-splines that are low-rank smoothers perform well with tensor product interactions and are known to work well with any penalty order and any combination of basis functions (Wood, 2017). For a detailed discussion on smoothing splines see Hastie and Tibshirani (1990), Ruppert et al. (2003), James et al. (2015) and Wood (2017) among others. The parameter estimates of Equation (10.2) are obtained by minimizing the function as follows:

$$Q_{Y|X}(\tau) = \sum_{t=1}^{n} \rho_\tau\left(y_{t,\tau} - \sum_{j=1}^{p} s_{j,\tau}\left(x_{tj}\right)\right) \tag{10.3}$$

where

$$\rho_\tau(v) = \left\{\tau 1(v \geq 0) + (1-\tau)1(v < 0)\right\}|v| \tag{10.4}$$

is the quantile loss function also known as the pinball loss function (PLF) and $1(.)$ denotes an indicator function. Combining Equations (10.1) and (10.2), Equation (10.3) reduces to

$$Q_{Y|X}(\tau) = \sum_{t=1}^{n} \rho_\tau \left(y_{t,\tau} - \sum_{j=1}^{p} \sum_{r=1}^{q} \beta_{rj} b_{rj}(x_{tj}) \right) \tag{10.5}$$

In the time series data that exhibit seasonalities, interactions are known to play a very important part in a regression-based model (Xie et al., 2016; Laurinec, 2017; among others). There are several ways of including interactions in AQR models, which include product of two predictor variables, such as $x_{tj} \times x_{tk}$, or $s(x_{tj}) \times x_{tk}$ or $s(x_{tj}) \times s(x_{tk})$ normally written as $s(x_{tj}, x_{tk})$, including more complex ways of representing interactions through tensor product interactions (Wood, 2006; Laurinec, 2017). Including pairwise interactions reduces Equation (10.1) to

$$y_{t,\tau} = \sum_{j=1}^{p} s_{j,\tau}(x_{tj}) + \sum_{k=1}^{K} \sum_{j=1}^{J} \tau_{jk} s_j(x_{tj}) s_k(x_{tk}) + \varepsilon_{t,\tau}, \tau \in (0,1) \tag{10.6}$$

10.2.2 The Proposed AQR Model

The variables considered in this chapter are the response variable, $y_{t,\tau}$, which denotes Daily peak electricity demand(DPED) at the τth quantile, the calendar effects, that is, hourly (H_{th}), where $t = 1, \ldots, n$ and $h = 1, \ldots, 24$; day of the week (D_{td}), where $d = 1$ denotes Monday, $d = 2$ denotes Tuesday, up to $d = 7$ denoting Sunday; monthly (M_{tm}), where $m = 1$ denotes January, $m = 2$ denotes February, up to $m = 12$ denoting December; holiday (R_{td}) that takes value 1 when day d is a holiday and 0 otherwise, a nonlinear trend (noltrend) that we get by fitting smoothing splines, lagged demand effects, that is, $y_{t,\tau-1}$ (Lag1) and $y_{t,\tau-2}$ (Lag2), respectively, including temperature variables that were derived by splitting South Africa into interior (inland) and coastal thermal regions. The temperature variables used in the study are: average daily temperature, average minimum daily temperature, and average maximum daily temperature, denoted by aveDT, aveMinT, and aveMaxT, respectively, including the minimum interior temperature (minIT), the difference between average minimum of interior and coastal temperatures (DaveMinTIC), and the difference between average of average daily coastal and interior temperatures (DaveADICT) as discussed in Sigauke (2017).

The model is then given as

$$y_{t,\tau} = \sum_{j=1}^{p} s_{j,\tau}(x_{tj}) + \sum_{k=1}^{K} \sum_{j=1}^{J} \tau_{jk} s_j(x_{tj}) s_k(x_{tk}) + \varepsilon_{t,\tau} \tag{10.7}$$

$$\emptyset(B)\Phi(B^s)\varepsilon_{t,\tau} = \theta(B)\Theta(B^s)v_{t,\tau} \tag{10.8}$$

$$\Rightarrow \varnothing(B)\Phi(B^s)\left[y_{t,\tau} - \left\{\sum_{j=1}^{p} s_{j,\tau}(x_{tj}) + \sum_{k=1}^{K}\sum_{j=1}^{J} \tau_{jk} s_j(x_{tj}) s_k(x_{tk})\right\}\right] = \theta(B)\Theta(B^s) v_{t,\tau}$$

$$(10.9)$$

where $\varepsilon_{t,\tau}$ denotes residuals that are correlated. The residuals are assumed to follow a seasonal autoregressive moving average process with s denoting the seasonal length, and $v_{t,\tau}$ being a white noise process with mean, zero, and variance, σ_v^2. Selection of variables will be done using Lasso for hierarchical interactions discussed in Bien et al. (2013) and implemented in the R package "hierNet" (Bien and Tibshirani, 2015). The objective is to include an interaction when both variables are included in the model. This restriction known as the strong hierarchy constraint is discussed in detail in the works of Bien and Tibshirani (2015) and Lim and Hastie (2015).

10.2.3 Parameter Estimation

Parameters of the developed models are estimated using the R package, "qgam" developed by (Fasiolo et al., 2017). In a Bayesian quantile regression framework, the likelihood function, $\pi(y\,|\,\beta)$, which is used to update the prior, $\pi(\beta)$ does not exist due to the fact that quantile regression is based on the pinball loss (Bissiri et al., 2016; Fasiolo et al., 2017). As a result, a belief-updating framework is used (Bissiri et al., 2016; Fasiolo et al., 2017). The posterior distribution of the belief updating of $\pi(\beta)$ is given as follows (Bassiri et al., 2016; Fasiolo et al., 2017):

$$\pi(\beta\,|\,y) \propto \vartheta e^{-\vartheta \sum_{i=1}^{n} L(y_i,\beta)} \pi(\beta)$$

$$(10.10)$$

where $\vartheta > 0$ determines the relative weight of the loss, $L(y_i,\beta)$, and the prior, $\pi(\beta)$. The parameter, ϑ, is known as the learning rate (Fasiolo et al., 2017).

Letting $\vartheta = \dfrac{1}{\sigma}$, the posterior distribution given in Equation (10.10) leads to the following expression (Fasiolo et al., 2017):

$$\pi(\beta\,|\,y) \propto \prod_{i=1}^{n} \frac{\tau(1-\tau)}{\sigma} \exp\left\{-\rho_\tau\left(\frac{y_i - \mu}{\sigma}\right)\right\} \pi(\beta)$$

$$(10.11)$$

The density, $\pi_{AL}(y\,|\,\mu,\sigma,\tau) = \dfrac{\tau(1-\tau)}{\sigma} \exp\left\{-\rho_\tau\left(\dfrac{y_i - \mu}{\sigma}\right)\right\}$, is the asymmetric Laplace density, where μ, σ, and τ are the location, scale, and asymmetry parameters, respectively. In this study, we use the decomposed σ from the work of Fasiolo et al. (2017) given as follows:

$$\sigma(\mathbf{x}) = \sigma_0 \exp\left\{\sum_{j=1}^{m} s_j(\mathbf{x})\right\} \tag{10.12}$$

where σ_0 represents the reciprocal of the learning rate.

10.2.4 Modeling Residual Autocorrelation

Since the residuals of the hourly electricity demand models are autocorrelated, it is important that the autocorrelation is significantly reduced before the models are used for forecasting (Hyndman and Fan, 2010; Hyndman and Athanasopoulos, 2013; among others). To reduce the autocorrelation, we adopt a method presented in Hyndman and Athanasopoulos (2013) and discussed in detail by Marasinghe (2014) and Sigauke (2017). The procedure for modeling residual autocorrelation that possibly exists in the residuals of the models is as follows: Estimate the parameters in Equation (10.7), extract residuals, $\varepsilon_{t,\tau}$, and test for autocorrelation. If the residuals are autocorrelated, fit an appropriate seasonal autoregressive integrated moving average (SARIMA) model. The fitted values of the residuals of the SARIMA model are then subtracted from the response variable $y_{t,\tau}$ to get a new transformed response variable, $y^*_{t,\tau}$. The transformed response variable is then regressed on the independent variables. The process is repeated if the residuals are still autocorrelated until they are uncorrelated or the autocorrelation is significantly reduced as desired.

10.2.5 Forecasting Electricity Demand When Covariates Are Known in Advance

Data for January 2008–December 2011 are used for training, while the remaining data, January 2012–December 2013, are used for testing. Operational forecasts will be from January 1, 2014, to December 31, 2016. The additive quantile regression models are without interactions and with pairwise interactions. Model without interactions is given as

$$y_{t,\tau} = \beta_{0,\tau} + s_{1,\tau}(D_{td}) + s_{2,\tau}(M_{tm}) + s_{3,\tau}(\text{noltrend}) + s_{4,\tau}(\text{Lag1})$$

$$+ s_{5,\tau}(\text{Lag2}) + s_{6,\tau}(\text{aveAED}) + s_{7,\tau}(\text{aveMminED}) + s_{8,\tau}(\text{aveMinTI})$$

$$+ s_{9,\tau}(\text{DaveMinTIC}) + s_{10,\tau}(\text{DaveADICT}) + \varepsilon_{t,\tau}$$

$$\tag{10.13}$$

The interaction variables are $I_1 = (\text{Lag1}, D_{td})$ and $I_2 = (\text{Lag2}, \text{noltrend})$. Model with interactions is given as

$$y_{t,\tau} = \beta_{0,\tau} + s_{1,\tau}(D_{td}) + s_{2,\tau}(M_{tm}) + s_{3,\tau}(\text{noltrend}) + s_{4,\tau}(\text{Lag1})$$

$$+ s_{5,\tau}(\text{Lag2}) + s_{6,\tau}(\text{aveAED}) + s_{7,\tau}(\text{aveMinED}) + s_{8,\tau}(\text{aveMinTI})$$

$$+ s_{9,\tau}(\text{DaveMinTIC}) + s_{10,\tau}(\text{DaveADICT}) + s_{11,\tau}(I_1) + s_{12,\tau}(I_2) + \varepsilon_{t,\tau}$$

$$(10.14)$$

10.2.6 Forecasting Electricity Demand When Covariates Are Not Known in Advance (Operational Forecasts)

We discuss the forecasting of the covariates that we need to use for forecasting the response variable. For the testing data described in Section 10.2.5, we now assume that the data for the covariates are not given. The variables to be predicted during the testing period are Lag1, Lag2, noltrend, AED, minED, minTI, DminTCI, DADTCI, I_1, and I_2. The variables used in predicting the covariates were D_{td}, M_{tm}, and linear "trend" variable, including pairwise tensor product interactions of these variables. The bases used were P-splines and cyclic cubic splines. The models were GAMs, AQR models, and gradient boosting. The R packages used were "mgcv" developed by Wood (2017), "qgam" by Fasiolo et al. (2017), and "caret" by Kuhn et al. (2017).

10.2.7 Forecast Combinations and Error Measures

It is well established in the literature that we get more accurate forecasts by combining forecasts from individual models (Bates and Granger, 1969; Clemen, 1989; Devaine et al., 2012; Gaillard, 2015; Shaub and Ellis, 2017; among others). Several error measures for probabilistic forecasting, which include among others the continuous rank probability score (CRPS) and log score including the PLF, are discussed in the literature. In this study, we are going to use the PLF and the CRPS. The PLF is given as follows:

$$L\left(\hat{Q}_{Y|X}(\tau)\right) = \begin{cases} \tau\left(y_{t,\tau} - \hat{Q}_{Y|X}(\tau)\right) & \text{if } y_{t,\tau} \geq \hat{Q}_{Y|X}(\tau) \\ (1-\tau)\left(\hat{Q}_{Y|X}(\tau) - y_{t,\tau}\right) & \text{if } y_{t,\tau} < \hat{Q}_{Y|X}(\tau) \end{cases} \qquad (10.15)$$

where $\hat{Q}_{Y|X}(\tau)$ is the quantile forecast and $y_{t,\tau}$ is the observation used for forecast evaluation. The CRPS that assesses the predictive performance of a forecast, $y_{t,\tau}$, in terms of the predictive distribution, F, is given in Equation (10.16). The score, denoted by $S(F, y_{t,\tau})$, represents the penalty that has to be minimized.

$$\text{CRPS}(F, y_{t,\tau}) = \int_{-\infty}^{\infty} (F(\tau) - 1(y_{t,\tau} \leq \tau)^2 \, d\tau \qquad (10.16)$$

where $1(.)$ represents an indicator function that is 1 if $y \leq \tau$ and 0 otherwise. An alternative form is

$$\text{CRPS}(F, y_{t,\tau}) = \int_{0}^{1} QS_{\tau}(F^{-1}(\tau), y_{t,\tau}) d\tau \qquad (10.17)$$

where QS_{τ} is the quantile score given as

$$QS_{\tau}(F^{-1}(\tau), y_{t,\tau}) = 2(1[y_{t,\tau} \leq F^{-1}(\tau)] - \tau)(F^{-1}(\tau) - y_{t,\tau}) \qquad (10.18)$$

10.3 Empirical Results

10.3.1 Exploratory Data Analysis

Figure 10.1 presents a summary of the descriptive statistics of DPED. As shown by the skewness and kurtosis values in Figure 10.1, DPED data does not follow a normal distribution. This is confirmed by the Jarque–Bera test at the 5% level of significance.

Figure 10.2 shows a time series plot of DPED. The density, normal quantile-to-quantile (QQ), and box plots all show that DPED data is nonnormal.

10.3.2 Results and Discussions

A plot of DPED overlaid with a nonlinear trend (solid curve) for the period is given in Figure 10.3.

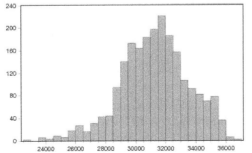

FIGURE 10.1
Summary Statistics of DPED (January 1, 2008, to December 31, 2013).

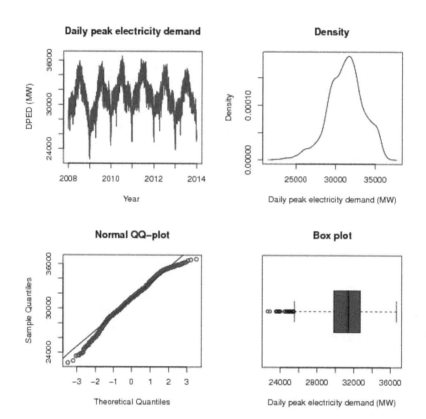

FIGURE 10.2
Daily peak electricity demand from January 1, 2008, to December 31, 2013.

The models developed are M1 (fqgam), M2 (fqgamI), and M3 (gbm). The "opera" R package developed by Gaillard et al. (2016) is used in combing the forecasts from the three models.

The weights assigned to the forecasts from the three methods are 0.0917, 0.717, and 0.191 for M1, M2, and M3 models, respectively. Figure 10.4 and Table 10.1 show the estimated pinball losses. Before combining the forecasts, model M2 is found to be the best model since it has the lowest pinball loss value. Combining the forecasts, the pinball loss of the convex model (M4) is the lowest.

A plot of actual DPED overlaid with combined forecasts for the testing period, January 1, 2012, to December 31, 2013, is given in Figure 10.5. The forecast values follow remarkably well with the actual demand.

An actual demand density plot overlaid with those from the combined and AQR models with interactions (fqgamI) is given in Figure 10.6.

The DPED density for the forecasted demand (January 1, 2012, to December 31, 2013) together with densities for quantiles $q_{0.01}$ and $q_{0.99}$ are shown in Figure 10.7. The $q_{0.99}$ shows that it is highly unlikely that DPED demand will

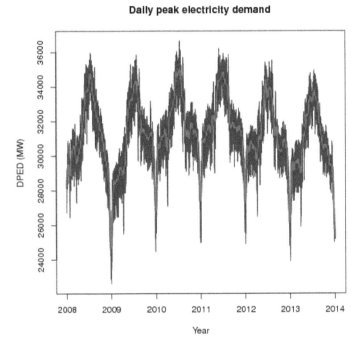

FIGURE 10.3
Plot of DPED overlaid a nonlinear trend (solid curve).

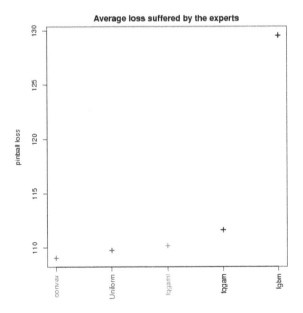

FIGURE 10.4
Plot of the experts (models) against the pinball loss.

TABLE 10.1

Error Measures (Forecast Given Covariates: January 1, 2012, to December 31, 2013)

	M1 (fqgam)	M2 (fqgamI)	M3 (fgbm)	M4 (Combined)
Pinball loss	111.6279	110.1792	129.5315	108.5632
CRPS	1152.902	1152.313	1156.65	1151.936

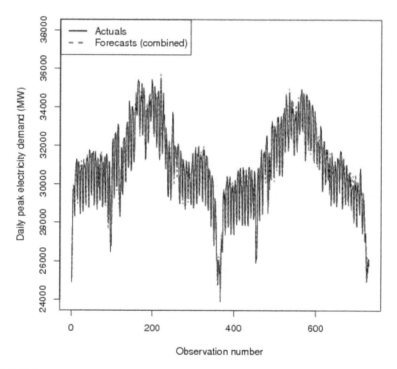

FIGURE 10.5
Actual demand overlaid with combined forecasts (January 1, 2012, to December 31, 2013).

exceed 38,000 MW, and the $q_{0.01}$ shows that it is highly unlikely for DPED to be below 23,500 MW. This is important for system operators to plan for the scheduling and dispatching of electricity to customers and for strategic managers to plan for capacity expansion.

10.3.3 Results: Operational Forecasts

On the basis of the pinball loss, model M5 (fqgamOS) has the smallest value of 457.0219, as shown in Table 10.2, confirming that it is the best fitting model. M5 is the AQR model without interactions.

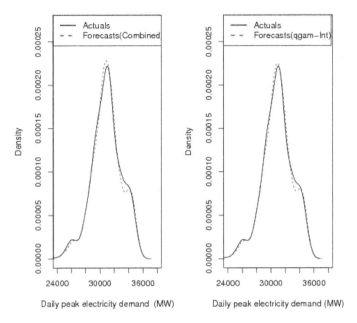

FIGURE 10.6
Actual demand density overlaid with densities from combined and AQR models with interactions (January 1, 2012, to December 31, 2013).

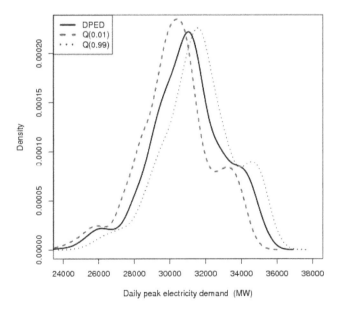

FIGURE 10.7
DPED density for the forecasted demand (January 1, 2012, to December 31, 2013) with densities for quantiles $q_{0.01}$ and $q_{0.99}$.

TABLE 10.2

Error Measures (Operational Forecasts: January 1, 2012, to December 31, 2013)

	M5 (fqgamOS)	M6 (fqgamOSI)	M7 (fgbmOS)
Pinball loss	457.0219	458.1457	458.3564

10.4 Point Process Characterization of the Occurrence of Extreme Peak Electricity Demand

10.4.1 Nonlinear Detrending: Modeling Extreme Peak Electricity Demand (Operational January 1, 2012, to December 31, 2016)

The DPED data is initially detrended by using a nonlinear detrending model. The model used in this study is a penalized cubic smoothing spline given as follows:

$$y_t = \sum_{t=1}^{n}\left(y_t - f(t)\right)^2 + \lambda \int \left(f''(t)\right)^2 dt + \varepsilon_t \tag{10.19}$$

and

$$\varepsilon_t = y_t - \hat{y}_t.$$

After fitting the model given in Equation (10.19) to the data, we extract the residuals and fit a nonparametric boundary corrected kernel density extremal mixture model discussed in Section 10.4.2.

10.4.2 Estimation of Threshold

We determine the threshold, τ, by fitting a boundary corrected extremal mixture model, discussed in MacDonald et al. (2013), on the positive residuals extracted from the fitted nonlinear detrending model discussed in Section 10.4.1. DPED observations above the threshold are then defined as extreme peak loads. Equation (10.20) shows the cumulative distribution function (CDF) of the boundary corrected extremal mixture model developed by MacDonald et al. (2013).

$$F\left(x|X,\tau,\gamma_{BC},\sigma_\tau,\xi,\theta_\tau\right)=\begin{cases}\left(1-\theta_\tau\right)\dfrac{H_{BC}\left(x|X,\gamma_{BC}\right)}{H_{BC}\left(\tau|X,\gamma_{BC}\right)}, & x\le\tau \\ \left(1-\theta_\tau\right)+\theta_\tau G\left(x|\tau,\sigma_\tau,\xi\right), & x>\tau\end{cases} \tag{10.20}$$

where τ is the threshold, σ_τ the scale parameter, ξ denotes the shape parameter, θ_τ is the exceedance probability, $H_{BC}(.|.)$ the CDF of the kernel density estimator, with the bandwidth represented by γ_{BC}, and $G(.|.)$ represents the generalized Pareto distribution (GPD) model fitted to observations above the threshold. The maximum likelihood method is then used to estimate the parameters. We use forecasts from model M5 (fqgamOS) (Figures 10.8 and 10.9).

We model the excesses ε_t and fit a point process above $\hat{\tau} = 1260.616\text{MW}$. Estimate intensity and frequency as follows:

$$\hat{\lambda} = \left[1 + \hat{\xi}\left(\frac{\hat{\tau} - \hat{\mu}}{\hat{\sigma}}\right)\right]^{-\frac{1}{\hat{\xi}}} \tag{10.21}$$

Figure 10.10 shows the estimated threshold ($\hat{\tau} = 1260.616\text{ MW}$), which is a vertical line separating the bulk model (kernel density) from the tail model (GPD).

With an estimated threshold of $\hat{\tau} = 1260.616\text{ MW}$ and the original series (positive residuals) of 1,071 observations, we got 104 exceedances. This resulted in an extremal index of 0.5074 using Ferro and Segers (2003) intervals estimator method (Figures 10.11 and 10.12).

The parameter estimates of μ, σ, and ε and their corresponding standard errors in parentheses are $\hat{\mu} = 1829.9(32.10)$, $\hat{\sigma} = 78.22(15.66)$, and $\hat{\xi} = -0.3616092(0.08566146)$, respectively, with estimated threshold $= 1260.616\text{ MW}$.

The estimated frequency using Equation (10.21) is given as follows:

$$\hat{\lambda} = \left[1 + (-0.3616092)\left(\frac{1261 - 1829.9}{78.22}\right)\right]^{-\frac{1}{(-0.3616092)}} = 35.34912279 \approx 35$$

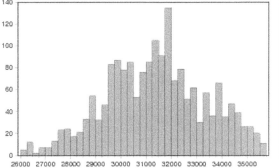

FIGURE 10.8
Summary statistics for operational forecasts of DPED (January 1, 2012, to December 31, 2016).

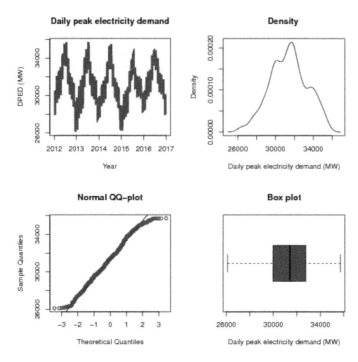

FIGURE 10.9
Operational forecasts of DPED (January 1, 2012, to December 31, 2016).

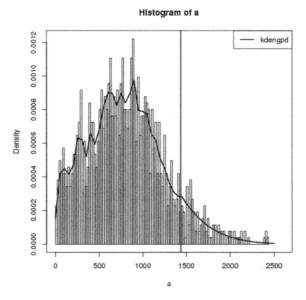

FIGURE 10.10
Threshold estimation ($\hat{\tau} = 1260.616$ MW).

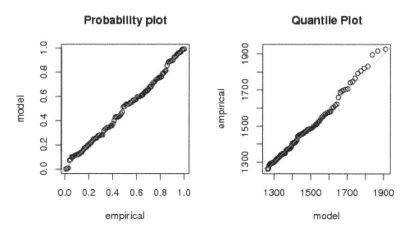

FIGURE 10.11
Diagnostic plots for the point process model fitted to exceedances above the threshold.

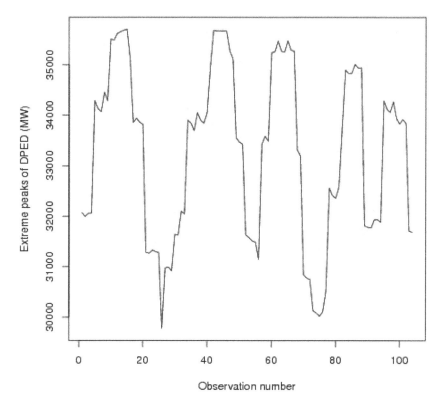

FIGURE 10.12
Extreme peaks of DPED for the period January 1, 2012, to December 31, 2016.

Using the orthogonal approach, we get

$$\hat{\lambda} = 365 \times \frac{\text{No. of } Y_i > 1260.616}{\text{No. of } Y_i} = 365 \times \frac{104}{1071} = 35.44351 \approx 35$$

This yields the same answer.

The forecasted extreme peak loads above 35,000 MW and the dates they are expected to occur on are given in Table 10.3.

Figure 10.13 shows that most daily peak loads occur between hours 18:00 and 20:00.

10.5 Reliability Analysis of Power Systems

It is important to have accurate and reliable DPED forecasts as they are useful in reliability assessments using indices such as generation reserve margin (GRM), loss of load probability, LOLE, and forced outage rates, including availability factors, among others (Čepin, 2011. We present two of some of the reliability indices, GRM and LOLE.

10.5.1 Reserve Margin

A reliability index that measures how a power system exceeds peak demand is referred to as the GRM and is calculated as (Čepin, 2011):

TABLE 10.3

Forecasted Extreme Peak Loads (January 1, 2012, to December 31, 2016)

Date	Monday, June 4, 2012	Monday, June 11, 2012	Monday, July 09, 2012	Monday, July 16,2012	Monday, July 23, 2012	Monday, July 30, 2012
DPED (MW)	35,509	35,490	35,623	35,658	35,687	35,706
Date	Tuesday, July 31, 2012	Monday, July 1, 2013	Monday, July 08, 2013	Monday, July 15, 2013	Monday, July 22, 2013	Monday, July 29, 2013
DPED (MW)	35,115	35,681	35,677	35,675	35,674	35,676
Date	Tuesday, July 30, 2013	Wednesday, July 31, 2013	Tuesday, June 03, 2014	Tuesday, June 10, 2014	Tuesday, July 22, 2014	Wednesday, July 23, 2014
DPED (MW)	35,273	35,117	35,244	35,256	35,474	35,270
Date	Thursday, July 24, 2014	Tuesday, July 29, 2014	Wednesday, July 30, 2014	Thursday, July 31, 2014	Tuesday, July 28, 2015	Tuesday, June 28, 2016
DPED (MW)	35,251	35,480	35,292	35,273	35,012	35,000

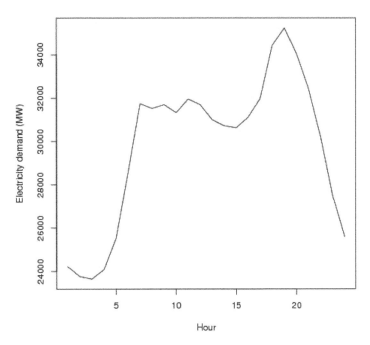

FIGURE 10.13
Typical daily load profile for South Africa (Thursday, June 13, 2013).

$$GRM = \frac{\text{capacity in service} - \text{peak load}}{\text{capacity in service}} \times 100$$

We shall use the GRM of 40,036 MW as discussed in Eskom Power Stations (n.d.). In 2012, the peak demand was 35,706 MW and it was on Monday July 20; in 2013, it was 35,681 MW and it was on Monday July 1; and in 2013, it was 35,480 MW and it was on Tuesday July 29. The calculated GRMs are given in Table 10.4.

International best practice generally requires a reserve margin that is at least 15% (Lam et al., 2016). Although there is a general increase in the reserve

TABLE 10.4

Estimated Reserve Margins

Year	Capacity (MW)	Peak Demand (MW)	Reserve Margin (%)
2012	40,036	35,706	10.82
2013	40,036	35,681	10.88
2014	40,036	35,480	11.38
2015	40,036	35,012	12.55
2016	40,036	35,000	12.58

margins for the forecasted peak DPED for the 5-year period 2012–2016 as shown in Table 10.4, all the predicted reserve margins are below 15%.

10.5.2 Loss of Load Expectation

Loss of load expectation, which is in days using daily peak load, is given as $\text{LOLE} = \sum_{i=1}^{n} p_i t_i$, where p_i is the probability of loss of load (loss of capacity in an outage) on day i and t_i denotes the duration of loss of power supply in days (Čepin, 2011). Outage risk as defined by Shenoy and Gorinevsky (2014) is the probability of outages that will occur in a year, which is given as $r = \dfrac{n_{out}}{N}$, where n_{out} is the number of outages in $N = 365$ days. Now assuming $n_{out} = 1$, that is, there is one outage per year, then $r = \dfrac{1}{365} = 0.00274$.

Equation (10.22) discussed in Čepin (2011) shows a formula used to calculate the expected number of days in a specified period in which the daily peak load exceeds the available capacity.

$$\text{LOLE}_p = \sum_{i=1}^{n} p_i(C_i - \text{DPED}_i)\text{days/period} \qquad (10.22)$$

where LOLE_p is the value of LOLE in period p in n days, C_i is the available capacity on day i, DPED_i is the forecasted peak load on day i, and p_i is the probability of loss of load on day i (Čepin, 2011). (Čepin, 2011) argues that "loss of load expectation method provides a consistent and sensitive measure of power generating system reliability".

10.5.3 Exponential Reliability Function

The Poisson process representing the number k of extreme peak load exceeding a predetermined threshold, τ in the time interval $(0,t)$ is given in the following equation (Chidodo and Lauria, 2012):

$$P_M(k,t) = P\big[N(t) = k\big] = e^{-\phi qt} \times \frac{(\phi qt)^k}{k!} \qquad (10.23)$$

where $N(t)$ is the arrival process that counts the number of extreme peak load exceeding the threshold $\tau = 1261$ MW over the period $(0,t)$, ϕ denotes the mean EPLF over $(0,t)$, and $q = 1 - F(\tau) = P(Y_i > \tau)$ is the exceedance probability. For the Poisson process $N(t)$,

$$E\big[N(t)\big] = \text{Var}\big[N(t)\big] = \phi qt = \phi t\big(1 - F(\tau)\big) \qquad (10.24)$$

The preceding equation holds under the assumption that the load process is stationary (Chidodo and Lauria, 2012). For the case $k = 0$, we get

$$P[N(t) = 0] = e^{-\phi qt} = e^{-\phi t(1-F(\tau))} \tag{10.25}$$

which represents the probability of no extreme peak load over $(0,t)$. This represents a "safety index" which is an exponential reliability function. With $\tau = 1261$, $t = 1$, $q = 0.09710551$, $\phi = 35$, we get $P[N(t) = 0] = e^{-35 \times 1 \times 0.09710551} = e^{-3.39869} = 0.0334169$, where $t = 1$, 1 denoting 1 year. This implies that the chances are very low of not getting an extreme peak load per year.

10.5.4 Discrete-Time Markov Chain Analysis

A three-state discrete-time Markov Chain problem is proposed in the modeling frequency of occurrences of DPED. The three states are given as follows:

1. Extreme ($\tau > 1260.616$)
2. Increase ($0 \leq \tau \leq 1260.616$)
3. Decrease ($\tau < 0$)

The transition matrix and transition diagram are given in Table 10.5 and Figure 10.14, respectively.

10.6 Conclusion

Power systems reliability requires a probabilistic characterization of extreme peak loads, which results in severe stress to the system and cause problems to the grid. Accurate predictions of long-term electricity demand are therefore very important as such forecasts can be used in the timing and rate of occurrence of such extreme peak loads. This is particularly important in reliability modeling.

TABLE 10.5

Transition Matrix of the States

	Extreme	Increase	Decrease
Extreme	0.40707965	0.5309735	0.0619469
Increase	0.03344482	0.6948161	0.2717391
Decrease	0.03061224	0.3469388	0.6224490

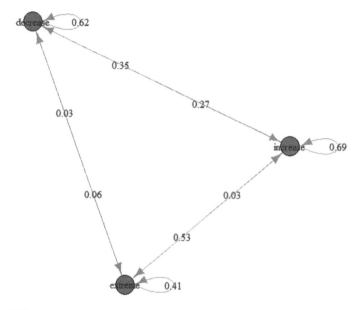

FIGURE 10.14
Transition diagram of the three states.

Acknowledgments

The authors thank Eskom, the power utility company of South Africa, and the South African Weather Services for providing electricity and temperature data, respectively. The study was funded by the National Research Foundation of South Africa (Grant No: 93613).

References

Bates, J.M. and Granger, C.W.J. (1969). The combination of forecasts. *Operational Research*, 20(4): 451–468.
Bien, J. and Tibshirani, R. (2015). R package "hierNet", version 1.6. https://cran.r-project.org/web/packages/hierNet/hierNet.pdf (Accessed 22 May 2017).
Bien, J., Taylor, J. and Tibshirani, R. (2013). A lasso for hierarchical interactions. *The Annals of Statistics*, 41(3): 1111–1141.
Bissiri, P.G., Holmes, C. and Walker, S.G. (2016). A general framework for updating belief distributions. *Journal of the Royal Statistical Society: Series B (Statistical Methodology)*, 78(5): 1103–1130.

Čepin, M. (2011). *Assessment of Power System Reliability: Methods and Applications.* Springer-Verlag, London.

Chidodo, E. and Lauria, D. (2012). Probabilistic description and prediction of electric peak power demand. *Electrical Systems for Aircraft, Railway and Ship Propulsion*: 1–7, DOI:10.1109/ESARS.2012.6387418.

Clemen, R.T. (1989). Combining forecasts: A review and annotated bibliography. *International Journal of Forecasting*, 5: 559–583.

Devaine, M., Gaillard, P., Goude, Y. and Stoltz, G. (2012). Forecasting the electricity consumption by aggregating specialized experts; a review of sequential aggregation of specialized experts, with an application to Slovakian and French country-wide one-day-ahead (half-) hourly predictions. *Machine Learning*, 90: 231–260.

Eskom Power Stations. (n.d.). http://www.eskom.co.za/OurCompany/PhotoGallery/Pages/Eskom_Power_Stations.aspx (Accessed 15 November 2017).

Fasiolo, M., Goude, Y., Nedellec, R. and Wood, S.N. (2017). Fast calibrated additive quantile regression. https://github.com/mfasiolo/qgam/blob/master/draft_qgam.pdf (Accessed 13 March 2017).

Fasiolo, M., Goude, Y., Nedellec, R. and Wood, S.N. (2017). Smooth additive quantile regression models, version 1.1.1 (Accessed 21 November 2017).

Ferro, C.A.T. and Segers, J. (2003). Inference for clusters of extreme values. *Journal of Royal Statistical Society Series B Statistical Methodology*, 65(2): 545–556.

Gaillard, P. (2015). Contributions to online robust aggregation: work on the approximation error and on probabilistic forecasting. PhD Thesis, University Paris-Sud.

Gaillard, P., Goude, Y. and Nedellec, R. (2016). Additive models and robust aggregation for GEFcom2014 probabilistic electric load and electricity price forecasting. *International Journal of Forecasting*, 32: 1038–1050.

Hastie, T. and Tibshirani, R. (1990). *Generalized Additive Models*. Chapman & Hall, London.

Hong, T. and Fan, S. (2016). Probabilistic electric load forecasting: A tutorial review. *International Journal of Forecasting*, 32: 914–938.

Hong, T., Pinson, P., Fan, S., Zareipour, H., Troccoli, A. and Hyndman, R.J. (2016). Probabilistic energy forecasting: Global Energy Forecasting Competition 2014 and beyond. *International Journal of Forecasting*, 32: 896–913.

Hyndman, R.J. and Fan, S. (2010). Density forecasting for long-term peak electricity demand. *IEEE Transactions on Power Systems*, 25(2): 1142–1153.

Hyndman, R.J. and Athanasopoulos, G. (2013). Forecasting: principles and practice, online textbook. http://otexts.com/fpp/.

James, G., Witten, D., Hastie, T. and Tibshirani, R. (2015). *An Introduction to Statistical Learning with Applications in R*. Springer, New York.

Kuhn, M., Wing, J., Weston, S., Williams, A., Keefer, C., Engelhardt, A., Cooper, T., Mayer, Z., Kenkel, B., Benesty, M., Lescarbeau, R., Ziem, A., Scrucca, L., Tang, Y., Candan, C. and Hunt, T. (2017). Classification and regression training, Version 6.0–76.

Lam, D., Sim, S., Steele, C. and Williams, J. (2016). FRCC 2016 load and resource reliability assessment reliability report, FRCC-MS-PL-081. Version 1. https://www.frcc.com/Reliability/Shared%20Documents/FRCC%20Reliability%20Assessments/FRCC%202016%20Load%20and%20Resource%20Reliability%20Assessment%20Report%20Approved%20070516.pdf.

Laurinec, P. (2017). Doing magic and analyzing seasonal time series with GAM (Generalized Additive Model) in R. https://petolau.github.io/Analyzing-double-seasonal-time-series-with-GAM-in-R/ (Accessed 23 February 2017).

Lim, M. and Hastie, T. (2015). Learning interactions via hierarchical group-lasso regularization. *Journal of Computational and Graphical Statistics*, 24(3): 627–654.

MacDonald, A.E., Scarrott, C.J. and Lee, D.S. (2013). Boundary correction, consistency and robustness of kernel densities using extreme value theory. Submitted.

Marasinghe, D. (2014). Quantile regression for climate data. All Theses. Paper 1909.

Maswanganyi, N., Sigauke, C. and Ranganai, E. (2017). Peak electricity demand forecasting using partially linear additive quantile regression models. *South African Statistical Journal: Peer-reviewed Proceedings of the 59th Annual Conference of the South African Statistical Association for 2017*, pp. 25–32. ISBN 978-1-86822-692-4.

McSharry, P.E., Bouwman, S. and Bloemhof, G. (2005). Probabilistic forecasts of the magnitude and timing of peak electricity demand. *IEEE Transactions on Power Systems*, 20(2): 1166–1172.

Ruppert, D., Wand, M.P. and Carroll, R.J. (2003). *Semiparametric Regression*. Cambridge University Press, Cambridge.

Shaub, D. and Ellis, P. (2017). Convenient functions for ensemble time series forecasts. R package version 1.1.9. https://cran.r-project.org/web/packages/forecastHybrid/forecastHybrid.pdf.

Shenoy, S. and Gorinevsky, D. (2014). Risk adjusted forecasting of electric power load. In *American Control Conference (ACC)*, pp. 914–919.

Sigauke, C. (2017). Forecasting medium-term electricity demand in a South African electric power supply system. *Journal of Energy in Southern Africa*, 28(4): 54–67.

Sigauke, C. and Bere, A. (2017). Modelling non-stationary time series using a peaks over threshold distribution with time varying covariates and threshold: An application to peak electricity demand. *Energy Journal*, 119: 152–166. ISSN: 0360–5442.

Wood, S. (2006). *Generalized Additive Models: An introduction with R*. Chapman and Hall, London.

Wood, S. (2017). P-splines with derivative-based penalties and tensor product smoothing of unevenly distributed data. *Statistics and Computing*, 27: 985–989.

Xie, J., Hong, T. and Kang, C. (2016). From high-resolution data to high-resolution probabilistic load forecasts. DOI:10.1109/TDC.2016.7520073. http://ieeexplore.ieee.org/document/7520073/ (Accessed 28 December 2016).

11

Real-Time Measurement and Evaluation as System Reliability Driver

Mario José Diván
National University of La Pampa

María Laura Sánchez Reynoso
National University of La Pampa

CONTENTS

11.1 Introduction 161
11.2 The Measurement as Key Asset 164
    11.2.1 C-INCAMI Framework 165
    11.2.2 Updating the Measurement Package 167
    11.2.3 CINCAMI/MIS: A Measurement Interchange Schema
            Based on C-INCAMI 169
11.3 The Processing Architecture Based on Measurement Metadata 172
11.4 The Organizational Memory in the Processing Architecture 174
11.5 Limiting In-Memory the Searching Space Related to the OM 176
    11.5.1 The Structural View: Filtering Based on the Common
            Attributes 177
    11.5.2 The Behavioral View: Filtering Based on the Processed Data ... 178
11.6 An Application Case: The Weather Radar of the National
      Institute of Agricultural Technology (Anguil) 182
11.7 Conclusions 184
References 185

11.1 Introduction

Along the decision-making process, the decision maker is unable to control everything and for this reason, there exists assumptions and premises among other considerations. The assumptions and premises are commonly associated with different aspects related to the context of a system, understanding the idea of system from the general system theory (von Bertalanffy and Hofkirchner, 2015). In the economy, there are many players with different sizes, each one has its own interest, and the key resides in adequately interpreting the particular interest and incorporating it in the big game (Cohn, 2016).

Information technology has a role of catalyzer in the economy, opening new ways and promoting continuously the use of data for knowing the current state of each aspect under monitoring (Pachenkov and Yashina, 2017). The data-driven decision making could be defined as the practice of basing decisions on the analysis of the data rather than purely intuition (Provost and Fawcett, 2013). This constitutes a key asset in real-time decision-making, because the analysis, and decisions should be made online, at the same time in which the data arrives from different data sources (Mohan and Potnis, 2017).

The phrase "Scrap In, Scrap Out" is applicable in analysis, data mining, and machine learning because if the quality of the data in the origin is doubtful, then the quality of the posterior analysis (the results) would be doubtful too (Suriyapriya et al., 2017). Each decision maker, at the moment in which he or she makes a decision, belives in their own data. In other words, when the decision maker uses the data for supporting different alternatives in its decision-making process, he considers that all the data are currents, consistents, traceables, comparables, and cohesives, among other characteristics. The potential risks associated with the bad data in the decision-making process have a direct effect on the expected results, which is dangerous in terms of the produced effects, for example, reputational damage and lost revenue, among others (Forbes Insights & Pitney Bowes, 2017).

The International Standard Organization (ISO) through the norm 25012 related to software engineering—Software Product Quality Requirements and Evaluation (SQuaRe)—establishes a multi-perspective analysis on the data characteristics (International Standard Organization, 2011). In this analysis, the standard proposes a way for characterizing the data and establishes the characteristics that are dependent on only the data itself or the system, or both.

In terms of the data quality model characteristics related to the ISO 25012 standard, the characteristics only dependent on the data are accuracy (syntactic and semantic), completeness, consistency, credibility, and currentness. The characteristics only dependent on the system are availability, portability, and recoverability. Additionally, the standard defines the characteristics jointly dependent on the data and the system at the same time, such as: accesibility, compliance, confidentiality, efficiency, precision, traceability, and understandability.

The proposed data characterization allows interpreting to the data quality as entity under analysis, so that each associated characteristic is plausible of quantification through different metrics. In this way and considering the measurement as quality driver, it is possible to define a measurement process for quantifying each data characteristic, and thus, it is likely to know whether the data gather or not the expected properties for being used along the decision-making process (Heinrich and Klier, 2015).

In real-time processing, it is highly likely that there exists different kinds of data sources (Kang et al., 2013). In this context, each sensor could send data in its own format. In terms of the sensor networks, the data gathering is a concern because the generated data arrive in eventually continuous ways, with fluctuating rates, and for that reason buffering and compression techniques are commonly used (Kishino et al., 2013). Thus, the data quality is a concern because on the received and processed data from the data sources, the real-time decision systems should make a decision on the fly and that could directly affect the system reliability (Mohan and Potnis, 2017).

Now, even when data-driven decision making is referred, any person could consider data and information as homogeneous concepts but would be fully erroneous. In fact, if information and data would be synonymous, then a decision maker could use any data from any place for supporting its decision-making process, but that would be very risky. The information could be defined as the data that satisfies simultaneously four properties: (i) consistency: the data has cohesivity and intern coherence; (ii) current: at least, the data represents the last known state; (iii) interest: they have particular interest for the decision maker and they are useful for the decision context; and (iv) opportunity: the data are timely and they are in the right moment when the decision maker must use them (Diván, 2012). For example, when a space shuttle is sent to the space, all the persons in the command center actively monitor all the indicators of the spacecraft for avoiding any kind of problem. In case of any problem, they possibly prefer to anticipate the situation through the data instead of solving a physical problem on the space shuttle, which would be truly riskier. In this context, the data arrives in real time (current), the data are of interest to the command center officer (interest), the data are gathered from truly calibrated data sources (consistent), and the data are fully available in the case in which a decision should be made (opportunity). In this scenario, the command center is able to make decisions on the basis of information. From this way, in real-time data processing oriented to supporting real-time decision making, it would be more appropiate to talk about information-driven decision making in place of just data-driven decision making.

However, data gathering in the sensor networks presents challenges in the moment in which the data should be collected because it is possible that each data source is different. In this sense, it is highly possible that the data schema, kind of data, precision, and accuracy, among other data source characteristics, are heterogeneous between the sensors of the network (Boubiche, 2016). This aspect is interesting in terms of system reliability, because it opens the possibility for incorporating a measurement interchange schema with the aim of homogenezing the measurement stream between each data source and the processor.

Now well, the idea related to homogenezing the data interchange between each heterogeneous data source and the real-time processor in a sensor network requires a previous agreement on the concepts and relationships involved in a measurement and evaluation (M&E) process. It is important because in a heterogeneous context with heterogeneous sensors, each sensor has its own way for storing the data, a scale, among other descriptive data on the measure. Each sensor collects data but doesn't know about the meaning of each measure with a given concept along the measurement process (Cullum Smith et al., 2014). It is possible for the processor to know the meaning of the value, but in a sensor network, the sensor could be heterogeneous, the number of sensors could be volatile, and each sensor could get a set of measures with different associated concepts and jointly send them at the same time.

An M&E framework allows to define the concepts, terms, and the neccesary relationships useful for formalyzing and implementing a measurement project. The frameworks agrees to the underlying concepts using some kind of domain model (e.g., an ontology), which fosters a common understanding among the set of sensors and the processors (Pesavento et al., 2017; Tonella et al., 2014). In this way, it is possible to establish a measurement interchange schema capitalizing the concepts defined in the framework for homogenizing the communication among the set of sensors and the real-time processor incrementing the system reliability (Cui et al., 2017).

The chapter is organized as follows: Section 11.2 introduces the Context-Information Need, Concept Model, Attribute, Metric and Indicator (C-INCAMI) M&E framework and highlights its importance along the measurement process jointly with its relationship with the data interchange schema. Section 11.3 synthesizes the processing architecture based on measurement metadata for remarking the importance among the C-INCAMI framework, the data interchange schema, the sensors, and the real-time processors. Section 11.4 presents the organizational memory as strategy for capitalizing the previous experiences in a structured way from the data interchange schema. Section 11.5 exposes a metadata-guided strategy for limiting in memory the search space related with the organizational memory before querying the big data repository with historical measures. Section 11.6 exposes an application case related to a weather radar (WR) located in the city of Anguil (Province of La Pampa, Argentina). Finally, Section 11.7 provides conclusions.

11.2 The Measurement as Key Asset

An M&E framework establishes the concepts, terms, and the necessary relationships for fostering a common understanding and makes it possible to implement a measurement process in a consistent way. A measurement project can thus be defined in terms of an M&E framework, and so their measures and analysis results can be consistently compared along the time in relation to its own definition (Mendonca and Basili, 2000). In this sense, what is the meaning of a metric? What is quantifying the metric? What is the scale and unit related to the metric´s measure? For example, if two different temperature sensors get the values in different scales and units, and the sensors send the data as raw data, then a conversion is mandatory for consistently comparing both values—only if the data processor is aware of this situation, else the comparison will be made on the basis of different values.

This aspect directly affects the system reliability; it is very important even in contexts such as sensor networks because the heterogeneity of each sensor or device represents a challenge in terms of the posterior data processing (Boubiche, 2016). That is to say, in real-time data processing the data itself should be processing on the fly, even when the communications are volatile. For example, the player satisfaction in real-time multiplayer mobile games is directly related to the performance of the communications network (Hansen et al., 2013).

Section 11.2.1 synthetically introduces the C-INCAMI framework and its importance in the M&E processes. Section 11.2.2 exposes the updates associated with the measurement package of C-INCAMI. This update is aligned with the application case related to the WR, which will be addressed in Section 11.6. Section 11.2.3 presents the measurement interchange schema developed from the updated C-INCAMI framework.

11.2.1 C-INCAMI Framework

C-INCAMI is a conceptual framework designed for supporting the M&E process in software organizations. Their terms, concepts, and relationships among them are defined through an ontological domain model (Olsina et al., 2007). Additionally, the framework incorporates the possibility of modeling the context related to the entity under analysis along the M&E process (Molina and Olsina, 2007).

In this approach, the requirement specifications, the M&E design, and the analysis of the associated results are oriented to satisfy a specific information need in a given context. Thus, a common understanding of data and metadata (e.g., the measurement project definition) is shared along the organization for fostering a consistent analysis (Becker et al., 2010).

The C-INCAMI frameworks is organized in six main components: (i) M&E project definition, (ii) nonfunctional requirements specification, (iii) context specification, (iv) measurement design and implementation, (v) evaluation design and implementation, and (vi) analysis and recommendation analysis. The M&E project definition (not shown in Figure 11.1) allows to define terms, concepts, and the needed relationships useful for dealing with the measurement activities, artifacts, and involved roles along the projects.

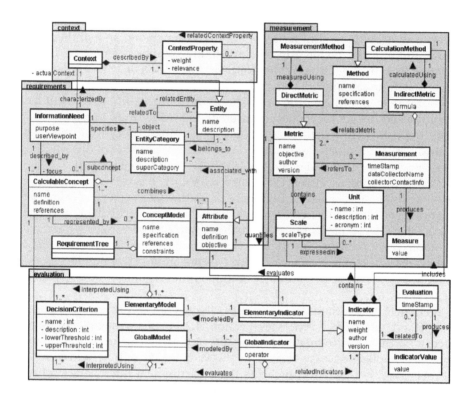

FIGURE 11.1
C-INCAMI main components and relationships for nonfunctional requirements specification, context specification, measurement design and implementation, and evaluation design and implementation components (Diván et al., 2011). This picture is reproduced with permission from Luis Olsina.

The *Nonfunctional Requirements Specification* (*requirement* package in Figure 11.1) allows to define the information need associated with the project. That is to say, it is possible to define the *purpose* (e.g., monitor) and the user *viewpoint* (e.g., data processor, etc.); in turn, it focuses on a *Calculable Concept* (e.g., system reliability) and specifies the *Entity Category* to evaluate (e.g., a resource). Each *Entity* (e.g., a system) under analysis belongs to an *Entity Category,* and the *Entity* could be characterized through *Attributes.* The Attributes could be quantified by metrics.

The *Context Specification* (*context* package in Figure 11.1) allows to describe the environment around the entity under analysis. In this sense, the *context* is a special kind of entity, and the *context properties* are a specialization of the attributes. In analogy with the attributes, the context properties allow characterizing the context related to the entity under analysis.

The *Measurement Design and Implementation* component (*measurement* package in Figure 11.1) establishes the concepts and relationships for designing and implementing the measurement process. A *Metric* establishes the measurement specification of how to quantify a particular attribute of an entity, using a *Method*, and how to represent its values, using a particular *Scale* in a given *Unit*. C-INCAMI proposes two kinds of metrics: direct and indirect. In the *Direct Metric*, the values are directly obtained applying the measurement method on the entity's attribute. Instead, the *Indirect Metric* obtains the values through a function and specific calculation method using other direct and indirect metrics as source. The concept *dataCollectorName* inside of the Measurement class (see Figure 11.1) represents the collector (e.g., sensor).

The *Evaluation Design and Implementation* Component (*evaluation* package in Figure 11.1) allows define the necessary concepts, terms and relationships for interpreting each value coming from a metric. That is to say, the metric gives a value, but the interpretation requires the specification of decision criteria for supporting the decision-making process. For this reason, this package allows defining the *Elementary Indicator*, which provides a mapping function between the measure's value to the indicator's scale (e.g., the level of energy consumed from a sensor). In this sense, a new scale is interpreted through the agreed decision criteria with the domain expert. The *Global Indicator* represents an aggregation from other global and elementary indicators, which makes possible global interpretation in light of the information need related to the project.

Finally, the *Analysis and Recommendation Specification* component (not shown in Figure 11.1) includes the concepts, terms, and relationships useful for the analysis, evaluation, and recommendation in an M&E project.

The C-INCAMI framework has an associated multipurpose strategy called GOCAME (Goal-Oriented Context-Aware Measurement and Evaluation). This strategy allows specifying the way in which an M&E project could be defined using the framework (Becker et al., 2011).

11.2.2 Updating the Measurement Package

The original C-INCAMI framework contemplates the essential details related to measurement package for the measurement process (see Figure 11.1). The proposed extension associated with the measurement component (Diván and Martín, 2017) allowed incorporating the aspects such as likelihood distributions in the measurement, among others, which are introduced in this section.

Figure 11.2 synthesizes the extensions related to the measurement package in C-INCAMI, and for better differentiation, the original concepts have the painted background, while new concepts have the white background associated.

The *QuantitativeMeasure* class inherits from the *Measure* class and it is associated exclusively with numerical measures. However, the *DeterministicMeasure* class represents the deterministic measures, while the *LikelihoodDistribution* class is related to estimated measurement process. When an estimated

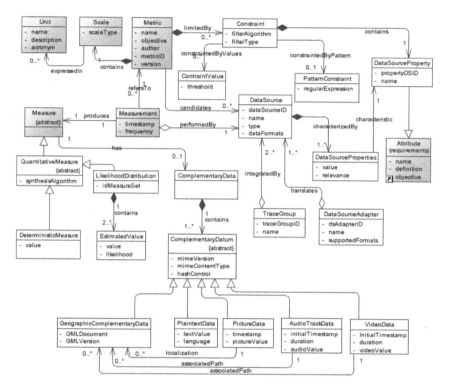

FIGURE 11.2

Extensions for the measurement components of C-INCAMI (Diván and Martín, 2017). This picture is reproduced with permission from Mario Diván and María de los Ángeles Martín.

measurement process returns the set of <value, likelihood> (see *Estimated Value* class), the *QuantitativeMeasure* class provides the *synthesisAlgorithm* property for computing a value from the set, following the directives given in the particular algorithm (e.g., a mathematical expectation). This is important because even when it is possible to get a given precision and accuracy, the related cost could be high, and a cheaper estimated process could be preferred (Tas et al., 2014; Castello Ferrer and Hanay 2015).

Now, each measure could have five different kinds of complementary datum (see the *ComplementaryData* and *ComplementaryDatum* classes) such as: geographic information, a plain text, a picture, a video, and an audio track. The complementary data contain a set of complementary datum. The associated idea is to gain contextual information related to the measurement context when it is necessary. As a highlighted aspect, the geographic information is organized in terms of Geography Markup Language (Open Geospatial Consortium and International Standard Organization, 2007) and a footprint is incorporated as property in the *ComplementaryDatum* class (see *hashControl* property) for controlling its associated integrity.

The *dataCollectorName* property related to the original *Measurement* class in C-INCAMI (see Figure 11.1) was replaced by a specific *DataSource* class (see Figure 11.2). The *DataSource* class represents the measurement device in the framework, and it incorporates four properties: (i) *dataSourceID*: it allows to identify the device along the measurement projects for fostering the traceability; (ii) *name*: a representative friendlier text for the device; and (iii) *type*: it refers to the way in which the sensor sends the data. It could be predictable when regularly sending data, or well unpredictable when it sends data in an irregular and volatile way, (iv) *dataFormats*: it refers to the schema or kind of organization related to the sent data from the sensor. Each data source sends the data through a data source adapter (see *DataSourceAdapter* class in Figure 11.2). The data source adapter represents the idea of gateway between the sensors and the real-time processor. It allows making the necessary conversions between the original format related to each sensor and a unified conceptual data schema such as the C-INCAMI/MIS schema (see Section 11.2.3).

The *TraceGroup* class represents the idea associated with the grouping of sensors. A data source could be described through the data source properties. Each data source property (see *DataSourceProperties* and *DataSourceProperty* classes in Figure 11.2) allows describing a particular characteristic related to the data source, indicating its relevance and the associated value (e.g., precision, accuracy, etc.).

In this sense, a metric should be implemented by a data source that satisfies all their associated minimum requirements. Thus, the *Constraint* class (see Figure 11.2) allows representing the minimum requirements imposed by the metric for its implementation (e.g., minimum accuracy).

11.2.3 CINCAMI/MIS: A Measurement Interchange Schema Based on C-INCAMI

One of the main characteristics related to the heterogeneous wireless sensor network (WSN) is its heterogeneity itself and the energy consumption that limits the lifetime of each sensor (Mahboub et al., 2017). The measurement interchange schema is an XML schema based on the extended C-INCAMI framework that allows capitalizing all their definitions, terms, concepts, and associated relationships. Thus, independent of the technology or kind of sensor, at the moment of the data gathering, the data are organized under CINCAMI/MIS, which allows homogenizing the provided data stream to the real-time data processor.

Figure 11.3 shows the top level of the measurement interchange schema. Three codes are used for describing the relations between tags along the XML hierarchy: (i) A: it represents that any number of child tags could be available under one father tag in any order; (ii) S: it represents that the tags under the father tag should be available in a specific order; and (iii) C: it represents that among the different tags under the father tag, just one of them must be chosen.

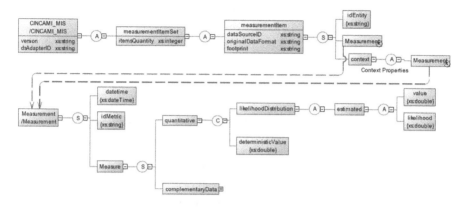

FIGURE 11.3

CINCAMI/measurement interchange schema (Diván and Martín, 2017). This picture is reproduced with permission from Mario Diván and María de los Ángeles Martín.

The top level of CINCAMI/MIS is related to the *CINCAMI_MIS* tag. This tag incorporates the *dsAdapterID* property, which allows to identify the sensor bridge, that is, the measurement adapter. In this sense, each CINCAMI/ MIS stream is associated with just one measurement adapter; however, it can incorporate measures coming from different sensors around it. This is important, because it is possible to make the differentiation between a failure related to the measurement device (e.g., sensor) and a failure associated with the data formatting.

The *measurementItemSet* tag under the *CINCAMI_MIS* tag could have any quantity of *MeasurementItem* tags. Each *MeasurementItem* tag has three properties: (i) *dataSourceID*: it identifies the origin of the data, that is, sensor; (ii) *originalDataFormat*: it is associated with the original data organization of the sensor; and (iii) *footprint*: it allows to verify the integrity of the measurement information under the *measurementItem* tag. Additionally, under the *MeasurementItem* tag, the entity under analysis shall be identified (*idEntity* tag), the associated measurement information is incorporated (*Measurement* tag) jointly with the quantification of its associated context.

The *Measurement* tag represents the measure (deterministic or not) and their associated complementary data. Recapitulating Section 11.2.1, the attributes and the context properties are quantified through the metrics and each metric has a particular objective (see Figure 11.1). In this way, through the *idMetric* tag under the *Measurement* tag, the real-time processor knows if the measurement corresponds to one specific characteristic of the entity under analysis or well if it is associated with a contextual property. The *datetime* tag represents the specific instant related to the measure and their complementary data, while the *Measure* tag is associated with the measure value itself.

A *measure* tag shall have the *quantitative* and *complementaryData* tags. The quantitative tag represents the measure, and if it is deterministic then the value is incorporated in the *deterministicValue* tag. However, if the measure is estimated, then the likelihood distribution is organized under the *likelihood-Distribution* tag as set of <Value, likelihood> (see *estimated* tag in Figure 11.3).

The *complementaryData* tag could have a set of *complementaryDatum* tag (see Figure 11.4). At the same time, it is possible to have different kinds of complementary data (e.g., a picture) as part of the complementary data. However, it is important to highlight that the *complementaryDatum* tag is optional. The *complementaryDatum* tag has three associated properties: (i) *mimeVersion*: it represents the mime version related to the complementary datum; (ii) *mimeContentType*: it describes the kind of content (e.g., image/jpeg); and (iii) *hashControl*: it is a footprint for verifying the content integrity (e.g., a MD5 footprint). The available kind of complementary datum for attaching under the *complementaryData* tag are as follows:

1. *GeographicComplementaryData*: it allows sending a document under the Geography Markup Language (Open Geospatial Consortium and International Standard Organization, 2007) as complement of the measure.

2. *PlainTextData*: it represents raw data coming from the sensor or related to the measure.

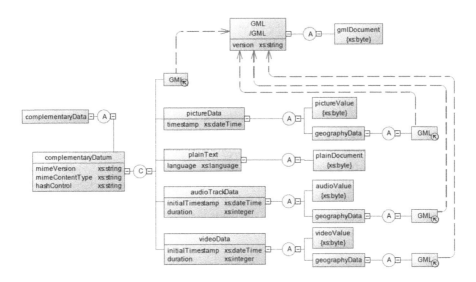

FIGURE 11.4
CINCAMI/measurement interchange schema. The complementary data organization (Diván and Martín, 2017). This modified picture is reproduced with permission from Mario Diván and María de los Ángeles Martín.

3. *PictureData*: it can incorporate a photo as complement of the measure.

4. *AudioTrackData*: it can incorporate an audio track as complement of the measure.

5. *VideoData*: it allows incorporating a video as complement of the measure.

The *PictureData*, *AudioTrackData*, and *VideoData* tags could have an associated GML document that represents its location as complement of the measure.

The measurement interchange schema allows to homogenize the data stream that comes from heterogeneous sensors, so that so the real-time processor makes the processing on the fly in a consistent way.

11.3 The Processing Architecture Based on Measurement Metadata

The processing architecture based on measurement metadata (PAbMM) is a semistructured data stream processor, which is specialized in M&E projects and it is implemented on a storm topology for real-time monitoring (Diván, 2016). PAbMM is based on the C-INCAMI framework for the definition of the M&E projects, and it is able to process C-INCAMI/MIS streams from the heterogeneous data sources.

Synthetically, PAbMM can be characterized by four processes: (i) *configuration*: it is responsible for defining the M&E project based on C-INCAMI framework, which will be implemented using the real-time processing architecture; (ii) *data collection and adaptation*: it is responsible for establishing the collecting methods and translating the raw data from heterogeneous data sources in a CINCAMI/MIS stream, implementing the load-shedding techniques (Rundensteiner et al., 2008); and (iii) *correction and analyses*: the smoothing and analysis functions are executed on the fly over each CINCAMI/MIS stream. These analyses inspect the stream based on the data and metadata inside the stream in correlation with the M&E project, (iv) *decision-making*: it is responsible for interpreting the decision criteria related to each indicator along the M&E project, for detecting and/or preventing abnormal situations. The decision making is able to use an organizational memory for capitalizing the previous experience. The processes were described using the SPEM (Software and Systems Process Engineering Metamodel) framework (Object Management Group (OMG), 2008) and information could be found in the work of Diván and Olsina (2014).

Before entering in detail with the conceptual architecture related to PAbMM, it is important to highlight that the M&E project definition is the first step before starting to try to monitor any entity (i.e., the configuration process).

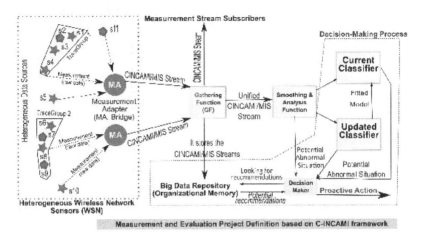

FIGURE 11.5
PAbMM. The conceptual architecture.

The architecture collects data from heterogeneous data sources (indicated as "s#" in Figure 11.5, where # represents the sensor ID) through the measurement adapter (MA) component (see MA in Figure 11.5). The sensors and MA are generally located along a heterogeneous WSN. The idea of MA was previously introduced in Sections 11.2.2 and 11.2.3. MA is basically a bridge between heterogeneous sensors and the gathering function (GF; see Figure 11.5). It is responsible for translating the raw data coming from a set of sensors in one unique CINCAMI/MIS stream.

The GF is responsible for gathering the CINCAMI/MIS streams from the measurement adapters, and unifing, storing, and replicating them in parallel. Thus, it is possible to synthesize each functionality as follows: (i) *the stream unification*: it is responsible for unifying all the measured characteristics of a given entity under a new and unified CINCAMI/MIS stream (see *Unified CINCAMI/ MIS stream* in Figure 11.5); (ii) *the stream storing*: each received CINCAMI/MIS stream from the MA is stored in the organizational memory for supporting future recommendations; and (iii) *the stream replication*: the architecture gives the possibility of consumption in real time and by subscription to the CINCAMI/MIS streams at the same time in which they arrive to the GF.

Once the unified CINCAMI/MIS arrives at the analysis and smoothing function (ASF; see Figure 11.5), this runs the principal component analysis, correlation analysis, and descriptive analysis. These analyses are guided by each tag inside the CINCAMI/MIS, which is related to the M&E project definition. If some abnormal situation is found, a descriptive message is sent to the decision maker.

Next, an online classifier based on Hoeffding Trees (Bifet et al., 2010) makes a decision based on the current situation, the associated indicator of the M&E project, and the entered CINCAMI/MIS stream (Current Classifier in Figure 11.5); if the decision corresponds with a possible abnormal situation, this is notified to the decision maker. In parallel, a new online classifier is obtained by the incorporation of the new CINCAMI/MIS stream and it makes a new decision; if this decision is related to a possible abnormal situation then it is communicated to the decision maker. A receiver operating characteristic curve (Marrocco et al., 2008) is used for comparing the classifiers, and if the updated classifier (UC) gets a gaining in relation to the current classifier (CC), then UC will replace CC as the new current classifier.

Finally, each time that the decision maker receives an abnormal situation notification, it analyzes if the situation is configured as disposable or not (e.g., the situation could be a tautology) along the M&E project definition. Thus, if the situation is configured as disposable, then it is immediately disposed. However, if the situation is not disposable, the decision maker looks for recommendations from the organizational memory for capitalizing the previous experience. In this way, the decision maker will make a decision and it will jointly send the associated recommendations for the particular situation.

Details about the storm topology associated with PAbMM can be found in the work of Diván and Martín (2017), and a data monetization strategy based on PAbMM was established in the work of Diván (2017).

11.4 The Organizational Memory in the Processing Architecture

Once the data has arrived at GF (see Figure 11.5), GF stores each CINCAMI/MIS stream in the Big Data repository, which implies that the data and metadata related to the measurement are stored together. The Organizational Memory (OM) serves as the basis for sharing the organizational knowledge, and it is useful for supporting the decision-making process (Barón, 2017).

The Big Data repository contains (i) the M&E project definition, (ii) the unclassified historical CINCAMI/MIS streams, and (iii) the OM.

The M&E project definition is key along the real-time processing system because all components, processing, and decisions are guided by the metadata that has their correlation in the M&E project definition.

The unclassified historical CINCAMI/MIS streams correspond with stored streams that are not associated with a known particular situation, be it abnormal or not.

The OM has each case in typified way with the descriptive information related to the CINCAMI/MIS stream, the known situations which

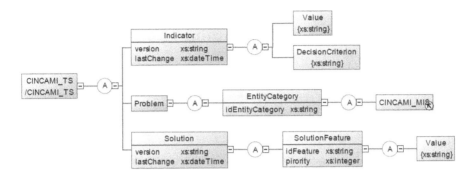

FIGURE 11.6
The CINCAMI/TS schema.

match, and their associated recommendations. For this reason, the OM uses the CINCAMI/TS (training set) schema as shown in Figure 11.6 for its structuration.

Basically, the CINCAMI/TS schema has three tags under the top tag called CINCAMI_TS: (i) *Indicator*: it represents the set of values associated with the decision criteria of an indicator for which the linked recommendations have sense (i.e., the proposed solutions); (ii) *Problem*: it allows describing the problem for a particular entity category. The associated behavior to the entity category is characterized through the measurements inside the chosen CINCAMI/MIS streams (see Figures 11.3 and 11.4). In this sense, the domain expert is able to identify and define the situations based on its knowledge and experience; and (iii) *Solution*: Each solution is characterized through the solution feature. Because each solution feature is not equal in terms of importance, each solution feature has its associated priority for the proposed solution.

Additionally, the organizational memory allows training the classifiers at the start-up. That is to say, even when the online classifiers are incremental, and the morphology is updated with each new CINCAMI/MIS stream in the processing architecture, at the beginning, they need minimal information for knowing about the kind of decision to be made.

A big data repository is characterized by the volume among other things (Singhal and Kunde, 2017) and the OM is not an exception. In this sense, if the decision maker (see Figure 11.5) at the moment in which an abnormal situation is notified wishes to look for recommendations, he or she should make a search along the big data repository, which has a high operational cost. Thus, and for optimizing the search along the OM, the next section introduces an in-memory filtering strategy based on the metadata stored in the M&E project definition, which allows limiting the searching space before the decision maker effectively reads data from the OM repository.

11.5 Limiting In-Memory the Searching Space Related to the OM

An entity and the context are characterized through the descriptive attributes and the contextual properties, respectively, based on the terms, concepts, and relationships established along the C-INCAMI framework (see Sections 11.2.1 and 11.2.2). These characterizations are defined at the moment in which the M&E project is created. It is important because all the components along the processing architecture (see Section 11.3) are then configured using the project definition. In this way, the data and metadata jointly arrive inside the CINCAMI/MIS stream (see Section 11.2.3) as a result of the M&E project definition.

In the previous section, the OM was introduced as a tool for supporting the decision-making process through the recommendations (see Section 11.4). PAbMM is a real-time processing architecture, while the OM is inside of a big data repository, and the data access times are very different between both. When the recommendations are necessaries, a logical idea would be to avoid unnecessary actions due to the cost associated with the I/O operations and to directly go to a specific region related to the big data repository. However, it is highly possible that not all the situations are previously defined by experts for a given M&E project. Thus, the challenge is related to how to proceed when it is important to look for recommendations for serving the decision maker (see Section 11.3); the search operation is also known to be expensive along the persistent big data repository.

In this sense, PAbMM proposes the possibility for filtering in-memory based on the project definition but differentiating between two views: (i) *the structural view*: it allows looking for structurally similar entities along all the M&E projects, generating a scoring from the most similar to the less similar entity; (ii) *the behavioral view*: using the structural scoring related to the entity, the real-time data are analyzed from the statistical point of view, for computing a correlation among the most similar entities but based on the behavior related to each common attribute among them. In this way, if the current M&E project definition does not have associated recommendations, the search is previously guided in-memory in PAbMM using the behavioral view based on the structural view. Next, PAbMM goes to the specific locations associated with the most similar entities for which it is assumed that the likelihood of finding similar situations is high. Given two entities, this assumption is based on (i) the attributes defined along the C-INCAMI framework that could be reused in different entities under monitoring, (ii) two entities that have common attributes described on the basis of the C-INCAMI framework, when they share at least one attribute, (iii) a similar behavior on a common attribute between two entities, which exposes a similarity grade between two situations.

Section 11.5.1 introduces details associated with the structural view for filtering in-memory, and Section 11.5.2 synthesizes the behavioral point of view.

11.5.1 The Structural View: Filtering Based on the Common Attributes

Each M&E project definition is kept in-memory by PAbMM because it is used for supervising the real-time data. So, the E set represents entities under monitoring, while e_1 and e_2 are two entities belonging to the E set.

The A set is associated with all available attributes in C-INCAMI, which makes it possible to describe a given entity category (see Figure 11.1). Thus, an entity is characterized for a variable quantity of attributes belonging to the set A.

The shared attributes between two entities under monitoring are defined as a natural number (N_0^+) and it is represented as $|e_1 \cap e_2|$. The expression $|e_1 \cup e_2|$ represents the quantity of different attributes associated with the entities e_1 and e_2. Because two entities do not share necessarily all the attributes, the *structural similarity coefficient* between two entities e_1 and e_2 is defined as:

$$sim\_str(e_1, e_2) = |e_1 \cap e_2| / |e_1 \cup e_2| \tag{11.1}$$

As was shown in Section 11.2.1, an entity is described by attributes and for that description, at least one attribute is required. So, the divisor related to the Equation (11.1) always will be greater than zero. Thus, the dividend will be a number between 0 (not-common attributes) and $|e_1 \cup e_2|$ (all the attributes are shared). In this way, the structural similarity is represented by number between 0 and 1, in which 0 represents no-common attributes, and 1 represents fully common attributes.

Now, using Equation (11.1) and based on C-INCAMI framework for the M&E project definition, it is possible to know whether two entities are structurally similar or not. Then, let $|E| = n$, a natural number that represents the quantity of entities under analysis in PAbMM, e_i and e_j entities, which belongs to E; the *structural similarity matrix* (SSM) is computed as follows:

$$SSM = \sum_{i=1...n} \sum_{j=(i+1)...n} sim\_str(e_i, e_j) \tag{11.2}$$

The computing happens for $i \neq j$ in Equation (11.2), because an entity is always exactly equal to itself and for that reason the main diagonal will be 1. Moreover, the resulting matrix (i.e., SSM) is triangular, because $sim\_str(e_i, e_j) = sim\_str(e_j, e_i)$.

In this sense and for given entity (e.g., e_3), it is possible to locate the route from the most structural similar entity to the less structural similar entity using

TABLE 11.1

A Conceptual Example Related to SSM

	e_1	e_2	e_3	...	e_n
e_1	1.00	0.90	0.80	...	0.70
e_2		1.00	0.90	...	0.95
e_3			1.00	...	0.85
...				1.00	...
e_n					1.00

Table 11.1: the route would be something like e_2 (0.90), e_n (0.85), ..., e_1 (0.80). The same procedure is used in-memory by the decision maker (see Section 11.3) for initially filtering the search space from the structural point of view when similar recommendations from other entities should be located.

Accordingly, this alternative does not warranty to always get recommendations through the OM because it is possible that such an experience does not exist yet. However, this mechanism allows extending the experiences of an entity for online recommendation, considering similar experiences at least from the structural point of view.

11.5.2 The Behavioral View: Filtering Based on the Processed Data

The second stage of filtering associated with the search space starts from the SSM and uses the statistical analysis made over each metric related to the attributes of the involved entities. Thus, this stage is known as the behavioral view because the analysis is now addressed over the data behavior associated with each attribute among the entities.

Retaking the structural similarity coefficient, on the one hand, when it is higher than zero, it implies that it has at least one common attribute (see Section 11.5.1). On the other hand, if the structural similarity coefficient is less than one, it implies that it has at least one different attribute between the entities. This is important to highlight because the behavioral analysis take sense when the structural similarity coefficient falls in the interval (0; 1]. Each metric has just one associated attribute or contextual property (see Sections 11.2.1 and 11.2.2).

In this way and so as to be able to quantify the relation between two quantitative metrics, it could use Spearman's correlation or even Kendall's correlation (James et al., 2017). Spearman's correlation is a nonparametric test that is used to measure the degree of association between two variables. Even when it does not have any supposition about the data distribution (e.g., normality), this is applicable when the variable is measured on a scale that is at least ordinal. In Spearman's test, the data should be monotonically related, which is an important consideration. Kendall's correlation is a nonparametric test that measures the strength of dependence between two variables.

Thus, given two entities named e_1 and e_2, the entity e_1 has $|e_1|$ attributes while the entity e_2 has $|e_2|$ attributes. In this way, it is possible to calculate Spearman's correlation by comparing the data of the metrics related to each entity's attribute. The expression Ma_{ij} represents the metric associated with the attribute 'j' from the entity 'i.' In this way, a Spearman correlation matrix could be computing among the metrics related to the entities e_1 and e_2, as shown in Table 11.2.

It is important to highlight that the correlation matrix does not give a number for associating with the concept of behavioral similarity; instead, the matrix allows Spearman's correlation along all the metrics related to each pair of attribute between two entities. For example, $S(Ma_{21}, Ma_{11})$ represents the correlation coefficient between the quantitative values associated with the metrics Ma_{21} (which quantifies the attribute 1 related to the entity 2) and Ma_{11} (which quantifies the attribute 1 related to the entity 1). This value is important because it allows to use the last available data from memory for knowing whether the metrics behavior data are similar.

$$\text{beh}_{\text{sim}(e_1,e_2)} = \sum_{i=1...|e_1|} \sum_{j=1...|e_2|} w_{1i} * w_{2j} * \sqrt{S(Ma_{1i}, Ma_{2j})^2} \qquad (11.3)$$

where

- w_{1i} represents weighing of the attribute i for the *entity* 1
- w_{2j} represents weighing of the attribute j for the *entity* 2
- $S(Ma_{1i}, Ma_{2j})$ represents the correlation coefficient between the values of the metrics Ma_{1i} and Ma_{2j}

In this way and through the computing of the $\text{beh}_{\text{sim}(e_1,e_2)}$ coefficient in Equation (11.3), it is possible to make a score in-memory for determining the behavior similarity among the entities with structural similarity. The scoring allows limiting the search space in previous ways to access the OM on the big data repository (see Sections 11.3 and 11.4).

TABLE 11.2

Spearman's Correlation Matrix between Entities

	Ma_{11}	Ma_{12}	Ma_{13}	...	$Ma_1|e_1|$
Ma_{21}	$S(Ma_{21}; Ma_{11})$	$S(Ma_{21}; Ma_{12})$	$S(Ma_{21}; Ma_{13})$	...	$S(Ma_{21}; Ma_1|e_1|)$
Ma_{22}	$S(Ma_{22}; Ma_{11})$	$S(Ma_{22}; Ma_{12})$	$S(Ma_{22}; Ma_{13})$	...	$S(Ma_{22}; Ma_1|e_1|)$
Ma_{23}	$S(Ma_{23}; Ma_{11})$	$S(Ma_{23}; Ma_{12})$	$S(Ma_{23}; Ma_{13})$	...	$S(Ma_{23}; Ma_1|e_1|)$
...				...	...
$Ma_2|e_2|$	$S(Ma_2|e_2|; Ma_{11})$	$S(Ma_2|e_2|; Ma_{12})$	$S(Ma_2|e_2|; Ma_{13})$	...	$S(Ma_2|e_2|; Ma_1|e_1|)$

The processing architecture uses R for statistical computing (R Core Team, 2017), and for that reason the correlation coefficients are computed through the R connection associated with the analysis and smoothing function.

The application of the behavioral similarity coefficient is quite simple, and to clarify, the theoretical case of three outpatients for which the monitoring is made along a heterogeneous sensors network is presented. All the outpatients correspond to the same Entity Category. For each outpatient, the experts have made a decision about the critical factors related to the vital signs for monitoring their health, to know: the corporal temperature, the systolic arterial pressure (maximum), the diastolic arterial pressure (minimum), and the cardiac frequency. Additionally, the experts could understand that the outpatient's context is important to supervise, and in that sense, the environmental temperature, the environmental pressure, the humidity, and the patient position are proposed as contextual properties (see Section 11.2.1).

In this case, the outpatients have a structural similarity of 1 because the monitoring happens on the same attributes and contextual properties related to each entity under analysis established in the M&E project definition. The point here is determining the behavior similarity among the outpatients in a given situation for limiting in-memory the searching space if some recommendations are necessary.

Thus, it is possible to define the associated variables as follows: (i) *cortemp*: corporal temperature, (ii) *sap_max*: the systolic arterial pressure—maximum, (iii) *dap_min*: the diastolic arterial pressure—minimum, and (iv) *cf*: the cardiac frequency. For simplicity, the weightings related to the entities are defined

```
> cor(outpatient1,outpatient2, method = "spearman")
              cortemp      sap_max       dap_min          cf
cortemp   0.44391582 -0.52784095   0.31331624  0.08918826
sap_max  -0.07909433  1.00000000  -0.03383325 -0.77639097
dap_min   0.41373566 -0.03383325   1.00000000 -0.39033070
cf        0.67764504 -0.46582439   0.22858180  0.05112473
> cor(outpatient1,outpatient3, method = "spearman")
              cortemp      sap_max       dap_min          cf
cortemp   0.5031153   0.1403122    0.31331624  0.08918826
sap_max   0.0000000  -0.1428571   -0.03383325 -0.77639097
dap_min   0.4387574   0.3721657    1.00000000 -0.39033070
cf        0.3504826  -0.1306989    0.22858180  0.05112473
> cor(outpatient2,outpatient3, method = "spearman")
              cortemp      sap_max       dap_min          cf
cortemp   0.4891200  -0.56053807   0.41373566 -0.2688655
sap_max   0.0000000  -0.14285714  -0.03383325 -0.7763910
dap_min   0.4387574   0.37216572   1.00000000 -0.3903307
cf       -0.0997155   0.04086268  -0.39033070  1.0000000
```

FIGURE 11.7
The Spearman's correlation using R Software.

in the same way as follows: *cortemp* = 0.2, *sap_max* = 0.25, *dap_min* = 0.25, and *cf* = 0.3. Each variable has an associated metric that allows quantifying the concept through some device, and the information will be sent through the CINCAMI/MIS stream jointly with the other variables (see Sections 11.2 and 11.3). Thus, a stream with ten measures related to outpatient 1, outpatient 2, and outpatient 3 is received, and for that reason, the correlation matrix using the Spearman method is computed (see Figure 11.7).

Figure 11.7 shows the correlation matrix organized as follows: outpatient 1 vs outpatient 2, outpatient 1 vs outpatient 3, and outpatient 2 vs outpatient 3. The correlation matrix computes the correlation between each metric between the two compared entities. From this information, it is possible to calculate the similarity behavior using Equation (11.3):

$$
\begin{aligned}
beh_{sim(outpatient1, outpatient2)} =& \left[(\mathbf{0.2*0.2}*0.4439) + (\mathbf{0.2*0.25}*0.5278) \right. \\
&+ (\mathbf{0.2*0.25}*0.3133) + (\mathbf{0.2*0.3}*0.0891) \left. \right] \\
&\left[(\mathbf{0.25*0.2}*0.079) + (\mathbf{0.25*0.25}*1) \right. \\
&+ (\mathbf{0.25*0.25}*0.0338) + (\mathbf{0.25*0.3}*0.7763) \left. \right] \\
&\left[(\mathbf{0.25*0.2}*0.4137) + (\mathbf{0.25*0.25}*0.0338) \right. \\
&+ (\mathbf{0.25*0.25}*1) + (\mathbf{0.25*0.3}*0.3903) \left. \right] \\
&\left[(\mathbf{0.3*0.2}*0.6776) + (\mathbf{0.3*0.25}*0.4658) \right. \\
&+ (\mathbf{0.3*0.25}*0.2285) + (\mathbf{0.3*0.3}*0.0511) \left. \right] \\
=& 0.4038
\end{aligned}
\tag{11.4}
$$

The bold text in the preceding equation is related to the weighting associated with each attribute or contextual property for the given entity. The italic font is associated with the correlation matrix between outpatient 1 and outpatient 2 (see Figure 11.7) for the given metrics. Thus, the behavioral similarity coefficient between outpatient 1 and 2 is 0.4038. By analogous reasoning for the pair outpatient 1 vs outpatient 3, it is possible to obtain its associated coefficient:

$$
beh_{sim(outpatient1, outpatient3)} = 0.3069
\tag{11.5}
$$

In this way and considering the results from Equations (11.4) and (11.5), the scoring of the more similar entities in relation to outpatient 1 is integrated by (i) outpatient 2 (0.4038) and (ii) outpatient 3 (0.3069). It is very important because when PAbMM does not have own recommendations for a specific situation of the outpatient 1, it will search along the OM following the order established in its scoring. If neither of them has valid recommendations for a given situation, PAbMM will send the alarm without recommendations.

The strategy based on the structural and behavioral coefficients allows avoiding any kind of unnecessary reads from the organizational memory, which represent a high cost in terms of I/O operations.

11.6 An Application Case: The Weather Radar of the National Institute of Agricultural Technology (Anguil)

The National Institute of Agricultural Technology (in Spanish, *Instituto Nacional de Tecnología Agropecuaria* [*INTA*]) have WRs located along the national territory for fostering the weather study among other applications. WRs are active sensors for remote sensing, which emit electromagnetic energy pulses in the atmosphere inside the microwave frequencies. The measurements are collected, on the one hand, through the way in which the electromagnetic radiation is propagated along the atmosphere and it is scattered by the objects and particles; on the other hand, the antenna capability for emitting directed radiation and to capture the radiation incident from a given direction. These sensors represent a valuable tool for monitoring the environmental variables, and they even allow the forecasting and evaluation of hydro-meteorological phenomena, among other possibilities.

The WR is basically composed of the following: (i) *the transmitter*: it allows generating a voltage signal, (ii) *the Antenna*: it allows giving a direction to the signal along the concentrated beam and also receives the radiation backscattered through the objects inside of the sampling volume, and (iii) *the receiving signal*: it includes the voltage induced in the antenna when it captures the incident radiation through which the processing is easier (Diván et al., 2015). The reflectivity is a measure related to the echoes intensity received by the radar antenna.

Figure 11.8a shows the WR infrastructure located in the city of Anguil, province of La Pampa in the middle of Argentina (South America). The radar is a Gematronic Meteor 600C model. The WR has a Doppler system and dual polarization, and it is able to operate along the C band (Gematronik, 2005). The antenna enables a shift in the horizontal direction (azimuth) and can rise up to 0.7854 radians vertical angle. This WR is set to complete a series of turns 2π radians that repeats on 12 elevation angles in ranges of 120 km, 240 km, and 480 km (Hartmann et al., 2010). An example of the first elevation is shown in Figure 11.8b. Each full scan is scheduled every 10 minutes.

The WR gets measures from each sample volume. Each measure is representative for the sample unit, which is 1 km^2 and 0.01745 radians (an example of precipitation accumulation [PAC] is available in Figure 11.8c). Among the

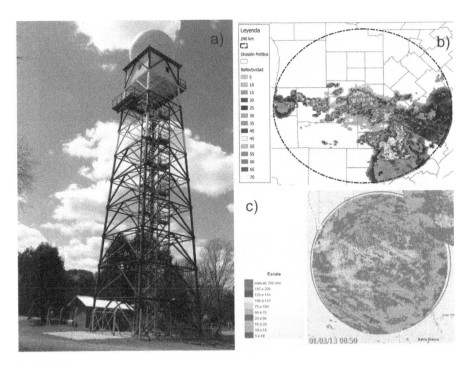

FIGURE 11.8
(a) INTA Anguil's WR infrastructure; (b) reflectivity image of the first elevation angle (0.00873 radians) on 15-01-2011 at 23:40 hours generated with INTA software; (c) the PAC of February 2013 generated with Rainbow 5 Gematronic software (Diván et al., 2015). This picture is reproduced with permission from Yanina Bellini (INTA EEA Anguil).

contained data along the measures, it is possible to find different variables: reflectivity factor (Z), differential reflectivity (ZDR), polarimetric correlation coefficient (RhoHV), differential phase (PhiDP), specific differential phase (KDP), radial velocity (V), and spectrum width (W) (Gematronik, 2005).

The WR is able to generate a minimum of 17 GB along a normal day without climate activity. This gives origin to a big data repository, because just in 1 year it will accumulate around of 6205 GB (365 days * 17 GB). The users of this kind of repositories are related to different industries from the meteorological services (government) to the assurance firms.

The original PAbMM was exclusively oriented to the real-time data processing (Diván et al., 2011). In Diván et al., 2015, an update of the PAbMM was raised, which incorporated the management of the big data repository and the real-time measurement replication. As shown in Section 11.4, this big data repository includes the M&E project definition, the unclassified historical CINCAMI/MIS streams, and the OM. The decision maker could be trained on the basis of different kinds of environmetal situations for sending online alarms and recommendations in each particular case. Additionally and because the metorological services in Argentina could require the data in real time, the PAbMM incorporates the idea of subscribers for immediately replicating the CINCAMI/MIS streams to them.

This Application case is completely based on the described architecture in Section 11.3. The technology associated with each component is as follows (see Figure 11.5):

- *Measurement Adapter and Gathering Function*: The Apacha Kafka (Apache Kafka Core Team, 2018) allowed replicating in real-time and in-parallel the streams to the subscribers, the big data repository, and to the Apache Storm Topology.

- *Real-Time Data Processing*: The GF, the smoothing and analysis function, the classifiers, and the decision maker were embedded inside of an Apache Storm Topology (Apache Storm Core Team, 2018).

- *The Statistical Computing*: The smoothing and analysis function was in-memory connected with the R software (R Core Team, 2017) for the statistical calculation. Additionally, Redis (Redis Labs Core Team, 2018) was used as in-memory database for storing temporal results along the computation.

- *The Big Data Repository*: It was implemented using Apache HBase (Apache HBase Core Team, 2018) as columnar database on Apache Hadoop (Apache Hadoop Core Team, 2018).

- *The Classifiers*: The classifiers used the Massive Online Analysis framework (Bifet et al., 2010) for implementing the Incremental Hoeffding Trees.

INTA has a set of heterogeneous WRs distributed along the Argentinian National Territory, and the PAbMM could be applicable for collecting data from each data source.

11.7 Conclusions

The chapter has introduced the relative importance of the measurement in relation to system reliability. Measurements are essential for knowing the quantitative value related to an entity under analysis. However, the value itself does not say much, it is in reality the evaluation that is responsible for interpreting each value considering the decision criteria defined by experts in the domain analysis. The M&E frameworks were presented as tools oriented to foster the repeatability, consistency, and comparability in the results along the measurement process. Thus, the M&E are essentially oriented to quantify an entity being monitored for knowing the current state and to be able to act in consequence. This aspect represents a key asset in relation to the system reliability.

Additionally, the processing architecture incorporates an OM for recommending courses of action in known situations. Even when this allows giving real-time alternatives to the decision maker for acting on the fly, the reading from big data repositories is expensive in terms of I/O operations.

However, and because not all the situations are previously known, the processing architecture incorporates a structural and behavioral filter for search recommendations. Thus, the possibility of filtering in-memory, looking for similar entities under analysis, it represents the possibility to limit the physical storage searching space in relation to the OM.

The application case related to the National Institute of Agricultural Technology allows schematizing the idea associated with the processing architecture in terms of the data sources, the processing, the replication, and the decision-making.

Many open work areas are currently in evolution: (i) the security aspects in relation to the confidentiality in data transmission, (ii) incorporation of a semantic similarity that will be able to detect attributes or contextual properties previously defined for avoiding redundancy, (iii) improving the search strategy on the OM guided by semantic similarity coefficients, and (iv) other aspects currently under research.

Nowadays, we cannot consider system reliability without measurement, and it is highly critical in real-time systems such as the data stream engines. The system reliability and the real-time M&E are two faces of the same coin; that is, the concepts of the system reliability and measurement have a symbiotic relationship and they are continuously in feedback.

References

Apache Hadoop Core Team. 2018. *Apache Hadoop*. Wakefield, MA: Apache Software Foundation.

Apache HBase Core Team. 2018. *Apache HBase*. Wakefield, MA: Apache Software Foundation.

Apache Kafka Core Team. 2018. *Apache Kafka. A Distributed Streaming Platform*. Wakefield, MA: Apache Software Foundation.

Apache Storm Core Team. 2018. *Apache Storm*. Wakefield, MA: Apache Software Foundation.

Barón, M. 2017. Applying social analysis for construction of organizational memory of R&D centers from lessons learned. *9th International Conference on Information Management and Engineering*. Barcelona: ACM. 217–220.

Becker, P, H Molina, and L Olsina. 2010. Measurement and evaluation as a quality driver. *Ingénierie de Systèmes d'Information (JISI)* 15(6): 33–62.

Becker, P, P Lew, and L Olsina. 2011. Strategy to improve quality for software applications: A process view. *Conference on Software and Systems Process*. Waikiki, Honolulu: ACM. 129–138.

von Bertalanffy, L, and W Hofkirchner. 2015. *General System Theory: Foundations, Development, Applications*. 1st ed. New York: George Braziller Inc.

Bifet, A, G Holmes, R Kirkby, and B Pfahringer. 2010. MOA: Massive online analysis. *The Journal of Machine Learning Research* (JMLR.org) 11: 1601–1604.

Boubiche, D. 2016. Secure and efficient big data gathering in heterogeneous wireless sensor networks. *International Conference on Internet of things and Cloud Computing*. Cambridge: ACM. 3:1–3:1.

Castello Ferrer, E, and S Hanay. 2015. Demo: A low-cost, highly customizable robotic platform for testing mobile sensor networks. *ACM International Symposium on Mobile Ad Hoc Networking and Computing*. Hangzhou: ACM. 411–412.

Cohn, T. 2016. *Global Political Economy: Theory and Practice*. 7th ed. New York: Routledge.

Cui, X, Y Peng, and X Peng. 2017. A universal test system framework and its application on satellite test. *International Conference on Compute and Data Analysis*. Lakeland, FL: ACM. 283–288.

Cullum Smith, G, J Hallstrom, S Esswein, G Eidson, and C Post. 2014. Managing metadata in heterogeneous sensor networks. *ACM Southeast Regional Conference*. Kennesaw, Georgia: ACM. 17:1–17:6.

Diván, M. 2012. *Information Technology Fundamentals for the Economics Sciences (In Spanish)*. Santa Rosa, La Pampa: EdUNLPam. Publishing House of the National University of La Pampa.

Diván, M. 2016. Processing architecture based on measurement metadata. *International Conference on Reliability, Infocom Technologies and Optimization (ICRITO)*. Noida: IEEE. 6–15.

Diván, M. 2017. Strategy for the data monetization in tune with the data stream processing. *International Conference on Reliability, Infocom Technologies and Optimization (ICRITO)*. Noida: IEEE. 78–88.

Diván, M, and L Olsina. 2014. Process view for a data stream processing strategy based on measurement metadata. *Electronic Journal of Informatics and Operations Research* 13(1): 16–34.

Diván, M, and M Martín. 2017. A new storm topology for synopsis management in the processing architecture. *XLIII Latin American Computer Conference (CLEI)*. Córdoba: IEEE. 1–10.

Diván, M, and M Martín. 2017. Towards a consistent measurement stream processing from heterogeneous data sources. *International Journal of Electrical and Computer Engineering (IJECE)* 7(6): 3164–3175.

Diván, M, L Olsina, and S Gordillo. 2011. Strategy for data stream processing based on measurement metadata: An outpatient monitoring scenario. *Journal of Software Engineering and Applications* (Scientific Research) 4: 653–665.

Diván, M, Y Bellini, M Martín, L Belmonte, G Lafuente, and J Caldera. 2015. Towards a data processing architecture for the weather radar of the INTA Anguil. *International Workshop on Data Mining with Industrial Applications (DMIA)*. San Lorenzo: IEEE. 72–78.

Forbes Insights & Pitney Bowes. 2017. Poor-quality data imposes costs and risks on businesses, Says New Forbes Insights Report. *Forbes Corporate Communication* (Forbes Insight).

Gematronik. 2005. *Rainbow® 5 Products & Algorithms*. Neuss: Gematronik GmbH.

Hansen, C, N Jurgens, D Makaroff, D Callele, and P Dueck. 2013. Network performance measurement framework for real-time multiplayer mobile games. *Annual Workshop on Network and Systems Support for Games*. Denver, CO: IEEE Press. 7:1–7:2.

Hartmann, T, M Tamburrino, and S Bareilles. 2010. Preliminar analysis of data obtained from the INTA radar network for the study of the precipitations in the pampeana region (in spanish). *39° Jornadas Argentinas de Informáticas –2° Congreso Argentino de Agroinformática*. Buenos Aires. 826.

Heinrich, B, and M Klier. 2015. Metric-based data quality assessment—Developing and evaluating a probability-based currency metric. *Decision Support Systems* 72: 82–96.

International Standard Organization. 2011. *BS/ISO IEC 25012. Software Engineering— Software Product Quality Requirements and Evaluation (SQuaRE)—Data Quality Model*. Standard: International Standard Organization.

James, J, D Witten, T Hastie, and R Tibshirani. 2017. *An Introduction to Statistical Learning with Applications in R*. 8th ed. New York: Springer Science+Business Media.

Kang, J, S Lee, H Kim, S Kim, D Culler, P Jung, T Choi, K Jo, and J Shim. 2013. High-fidelity environmental monitoring using wireless sensor networks. *Proceedings of the 11th ACM Conference on Embedded Networked Sensor Systems*. Roma: ACM. 67: 1–67:2.

Kishino, Y, Y Sakurai, Y Yanagisawa, T Suyama, and F Naya. 2013. SVD-based hierarchical data gathering for environmental monitoring. *ACM conference on Pervasive and ubiquitous computing adjunct publication*. Zurich: ACM. 9–12.

Mahboub, A, E En-Naimi, M Arioua, and H Barkouk. 2017. Distributed energy efficient clustering algorithm based on fuzzy logic approach applied for heterogeneous WSN. *2nd International Conference on Computing and Wireless Communication Systems*. Larache: ACM. 38:1–38:6.

Marrocco, C, R Duin, and F Tortorella. 2008. Maximizing the area under the ROC curve by pairwise feature combination. *ACM Pattern Recognition (ACM)* 41(6): 1961–1974.

Mendonca, M, and V Basili. 2000. Validation of an approach for improving existing measurement frameworks. *IEEE Transactions on Software Engineering (IEEE)* 26(6): 484–499.

Mohan, L, and D Potnis. 2017. Real-time decision-making to serve the unbanked poor in the developing world. *Proceedings of the 2017 ACM SIGMIS Conference on Computers and People Research*. Bangalore: ACM. 183–184.

Molina, H, and L Olsina. 2007. Towards the support of contextual information to a measurement and evaluation framework. *Quality of Information and Communications Technology (QUATIC)*. Lisbon: IEEE. 154–163.

Object Management Group (OMG). 2008. *Software and Systems Process Engineering Meta-Model Specification (SPEM)*. Standard: Object Management Group (OMG), Object Management Group (OMG).

Olsina, L, F Papa, and H Molina. 2007. How to measure and evaluate web applications in a consistent way. En *Web Engineering: Modelling and Implementing Web Applications*, editado por G Rossi, O Pastor, D Schwabe and L Olsina, 385–420. London: Springer-Verlag.

Open Geospatial Consortium and International Standard Organization. 2007. *ISO 19136:2007 Geographic Information—Geography Markup Language*. Standard: International Standard Organization.

Pachenkov, O, and A Yashina. 2017. When sharing economy meets digital one: towards understanding of new economic relations. *eGose '17: Proceedings of the Internationsl Conference on Electronic Governance and Open Society: Challenges in Eurasia*. St. Petersburg: ACM. 91–98.

Pesavento, D, O Mimouni, E Newberry, L Benmohamed, and A Battou. 2017. A network measurement framework for named data networks. *4th ACM Conference on Information-Centric Networking*. Berlin: ACM. 200–201.

Provost, F, and T Fawcett. 2013. Data science and its relationship to big data and data-driven decision making. *Big Data (Mary Ann Liebert, Inc.)* 1(1): 51–59.

R Core Team. 2017. *R: A Language and Environment for Statistical Computing*. Vienna: R Foundation for Statistical Computing.

Redis Labs Core Team. 2018. *Redis*. Mountain View, CA: Redis Labs.

Rundensteiner, W, M Mani, and M Wei. 2008. Utility-driven load shedding for XML stream processing. *International World Wide Web (WWW) Conference*. Beijing (China): ACM Press. 855–864.

Singhal, R, and S Kunde. 2017. Technology migration challenges in a big data architecture stack. *8th ACM/SPEC on International Conference on Performance Engineering*. L'Aquila: ACM. 159–160.

Suriyapriya, E, M Praveena, and M Phil. 2017. Clustering and Boosting in Data Mining. *International Journal of Engineering Science and Computing* 7(8): 14633–14634.

Tas, B, N Altiparmak, and A Tosun. 2014. Low-cost indoor location management for robots using IR leds and an IR camera. *ACM Transactions on Sensor Networks (TOSN) (ACM)* 10(4): 63: 1–63:41.

Tonella, P, M Ceccato, B De Sutter, and B Coppens. 2014. POSTER: A measurement framework to quantify software protections. *ACM SIGSAC Conference on Computer and Communications Security*. Scottsdale, Arizona: ACM. 1505–1507.

12

Optimizing Price, Release, and Testing Stop Time Decisions of a Software Product

A. K. Shrivastava

Fortune Institute of International Business

Subhrata Das, Adarsh Anand, and Ompal Singh

University of Delhi

CONTENTS

12.1 Introduction 189
12.2 Modeling Framework 194
    12.2.1 Notations 195
    12.2.2 Assumptions 195
    12.2.3 Proposed Sales Model 196
    12.2.4 Proposed Optimization Model 196
        12.2.4.1 Setup Cost 196
        12.2.4.2 Market Opportunity Cost 197
        12.2.4.3 Cost of Removal of Fault during Testing Phase
                before Release $[0, \tau]$ 197
        12.2.4.4 Cost of Removal of Fault during Testing Phase
                after Release $[\tau, T]$ 197
        12.2.4.5 Cost of Removal of Faults after Testing Stop Time T 198
        12.2.4.6 Cost of Testing 199
        12.2.4.7 Total Expected Software Cost 199
        12.2.4.8 Revenue from the Expected Sales is Given By 199
        12.2.4.9 Expected Profit 199
12.3 Numerical Illustration 200
12.4 Conclusion 201
Acknowledgments 202
References 202

12.1 Introduction

Software frameworks are considered to be an integral part of human life and have made every aspect of our everyday life dependent on it. Everything that we come across is somehow or the other way subjected to software system for its proper functioning. With worldwide progression in innovation it has turned out to be inescapable to maintain a strategic distance from software in our ordinary lives. Nowadays, major bulk of industries are getting more and more reliant on software systems, making themselves prone to several associated risks. It is very wise to say that around 50% of the world population utilizes web, and if continued with the same rate, this will soon become 100%. It is most likely that software frameworks have made our life simple; however, in the meantime, it has expanded the hazard on human lives and assets relying upon them. Due to the increased pace of software reliance, the market rivalry is at peak, and in turn, software development firms are amending their strategies to acquire the greatest advantage in the present situations. Influencing strategies keeping in mind the end goal of maintaining their position in the market at such a great position is an exceptionally extreme task today. Certain set of policies have been built that are robust and forms the foundation. Reliable and well-framed set of strategies can lead to an extraordinary pinnacle and at the same time will be able to maintain the trust of clients. Each organization experiences few circumstances in which they need to settle for ideal choices at different stages. These situations pertain to decisions taken during the development phase of the software, namely, for pricing, testing of the software, and also release time of the software. These choices have vital significance as they can fundamentally influence the objectives of software firms. The future of software is generally determined by the set of strategies that are adopted before and after its launch.

Testing phase is a standout among the most imperative stages in the software life cycle, which helps in concluding that the software application is now a deliverable item that can meet both designer and client prerequisites. Software testing brings up the imperfections that are implanted in the midst of the development cycle. It measures and upgrades the nature of software, which helps in building the customer feeling of confidence and satisfaction. Moreover, it helps drop down the expenses to upkeep both the client and designer coming about into more steady, reliable, and precise conclusions. No software product can be impeccable and subsequently without blunders; however, testing can limit the related hazard by diminishing the quantity of bugs. It is necessary to have a group of testers who can work for the organizations to outlive in changing market scenarios and setting up client confidence and trust in their software. This can only be achieved by a broad testing of the software that identifies, segregates, and then redresses the shortcomings present in it.

With a specific end goal to choose the software launch time, the administration is reliant upon its testing group that is in charge of discovering bugs in the software and their expulsion before it is used in the field. Bugs that are found amidst the testing stage can be utilized as thoughts for highlight upgrade. Subsequently, testing can end up being remedial and also a creative period of a product life cycle. Bugs found in the software during early phases of improvement lessen the cost of their removal. Consequently, it is essential to discover the blunders as ahead of schedule as could be expected under the critical circumstances.

In the current work on software testing, the ideal time to quit testing of a product corresponds with the ideal launch time of a product. In any case, this approach can be additionally enhanced by altering these two focuses with the end goal that product is launched early and testing proceeds after release till a particular unwavering quality level is accomplished. Early entry of software in the field can give higher advantages to the firm because of the chance of catching high market potential. Yet, on the other hand, because of less testing, more number of users reporting failures during operational phase which in turn will build the high cost of fixing, to the seller. Then again, late releases prompt utilization of more assets and thus increment the cost of developing the software. In addition, firms entering very late into the field result in losses due to missing market opportunities. Thus, deciding optimal release time has become a major concern for software developing firms as they are found liable in deciding the fortune of any software firm and are termed as software release time problems.

Figure 12.1 portrays the two different situations, one with no post-release testing and the other with post-release testing. In the later situation of testing after release, the firm probably plans to catch the market for a more drawn-out span of time, which in the long run adds up to more prominent advantages for the software developers. Dividing the testing stage into pre-release and post-release testing stages causes the developers to considerably expand their testing base from a set number of analyzers to a fundamentally huge number of analyzers during the shift from pre-release to post release testing phase of the software. In addition, an early release has upper hands over the competitors' offering. An early release implies early response, which forms the basis for the association and impacts the future objective of software developers. However, to maintain their threshold level, firms keep on testing the software and at the same time rolling the updates for enhancements of the features.

Increasing concern toward safety and inculcation of software into critical systems has become quite supreme and makes it crucial to think rationally about the decision pertaining to the launch time of the software. Many researchers have worked behind the ideology of obtaining an optimal release time of the software based on different combination of attributes (Pham, 2003, 2006; Musa, 2004; Kapur et al., 2011, Kapur et al., 2013). The optimization problems based on determination of release time of the software has to make a trade-off between releasing too early and testing for a longer duration. Early entry into the field increases the chances of failure, which in turn impact the quality of the software. Entering too late has a positive impact on the quality and reliability of the software, but the cost incurred due to a longer testing period shoots up (Software project, 2012; Continuous

Release Policies with and Without Postrelease Testing

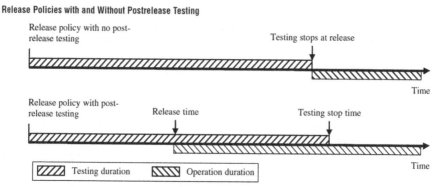

FIGURE 12.1

Release policy with and without post-release testing.

Delivery, 2013). Various examples can be quoted in which the failure of software system has resulted into huge losses (galorath.com), which depicts the essentialness by which developers should design the systems. A few current software crashes caused by latent bugs are as follows (Top 10 Mega, 2015):

- Screwfix endured a high anomaly because of which every single item on its site went available at just £34.99. Several clients appreciated and cherished the pleasure of having the same cost for every single product available on their website, which included products such as garden tractors to costly power devices. This resulted in Screwfix enduring an immense loss.

- In late 2014, during a festive bonanza offer, Amazon, being one of the largest online stores, confronted a massive loss of $100,000. This was due to some failure in the price comparison software due to which customers were able to choose and buy expensive items at a very low price.

- Failure in the tax collection system in the UK resulted in huge losses due to which more than 5.7 million people were affected for a period of 2 years. Due to the error in the tax computation system, those who were to pay less were asked to pay more and vice versa.

These reported incidents clearly define that at times small failures can also result into huge losses. A particular software can never be claimed to be fault free, but carrying the testing for a longer period and proper scheduling might decrease the chances of failures. In literature several ways to quantify the faults exist, which can be utilized to gauge the quantity of imperfections and their appropriation in software system (Musa, 2004; Pham, 2006; Kapur et al., 2011). They give a numerical relationship between different traits as testing advances. Different researchers contributed to the release time issue in light of the two basic qualities from the perspectives of the client and developers: cost and unwavering quality.

In 1980, Okumoto and Goel proposed a relatively simple software launch time optimization problem based on the arrangements in light of cost and unwavering quality. They proposed two different frameworks—the first one depends on an unconstrained cost objective compelled to maintain a threshold level of reliability and the other one is reliability maximization under with budgetary constraint. Yamada and Osaki (1987) inspected release time issues in view of concurrent cost minimization and reliability escalation objective, whereas the framework of Yun and Bai (1990) completely focused on a cost-streamlining model in light of arbitrary life cycle of the software. In 2003, Xie and Yang worked to obtain the optimal launch time under the concept of imperfect debugging, and Pham proposed cost models by consolidating several factors, namely, imperfect debugging, risk cost, warranty cost, penalty cost, and random life cycle. Huang (2005) and Huang and Lyu (2005) examined the release time decision problem thinking about the impact of testing proficiency and efforts consumed. Yang et al. (2008) proposed a cost model in light of cost vagueness and its effect on release time decision. Ukimoto and Dohi (2013) proposed a cost optimization model in light of two diverse operational situations with dependability imperatives.

The given set of examined studies has concentrated upon the combination of cost and reliability of the software and a major assumption that as soon as the software is given to users in the field it is no more tested. Evolution in the field of technology and intensified competition has resulted into an era where firms try to penetrate early into the field and keep on testing the software even after its launch to stay ahead of their competitor, but at the same time they wish to offer the quality product (HP, 2014). This policy has attracted several researchers to work upon and determine the optimal time of launch and also the time duration for which the software shall undergo testing subsequent to its release. The concept of releasing the software and continuing testing was basically introduced by Arora et al. (2006) and their concept was extended by Cavusoglu and Zhang (2008). They suggested that it is smarter to release early and fix the latent faults in the post-release period of testing. A cost modeling framework for optimal scheduling of updates for a particular software was worked upon by Luo et al. (2015). In line with the proposal of early software release in the year 2012, Jiang et al. proposed the cost frame giving consideration to market opportunity cost, which is defined as, "The opportunity to get higher revenue from the product by releasing the product in the time interval of early and actual release time." In their study it was claimed that it is economical for software firms to release early and engage in prolonged testing. Working on the similar grounds, Jiang et al. (2012) Kapur et al. (2015, 2017) built up a generic structure to obtain the launch time and testing stop time. In 2017, they extended their framework to evaluate the release and testing stop time of software under two constraints, namely, reliability and budget. Anand et al. (2017) also developed a cost model to see the impact of patching on software release and testing stop time decision and provided an optimal scheduling policy. Moving toward the field of utility theory, Majumdar et al. (2017) worked in developing a framework to distinguish release and testing stop time and obtained their optimal values using the multi-attribute utility theory. In the work done by Sachdeva et al. (2018), focus is laid on determining the time to stop testing and release time separately by considering warranty as an attribute. They were also able to differentiate their study from that of the others by revealing a time period in which both user and tester together focus on debugging faults from the software.

On the basis of the aforementioned studies, we assumed that software can be delivered prior to completion of testing. The above-discussed literature in the field of software engineering presents one of the significant shortcomings pertaining to pricing of software before its launch in the market. Pricing of software is dependent upon the resources utilized during software development process. Software pricing influences the buying behavior of the users for any particular software. Customers need good quality product at lower prices, but providing good quality product needs effort and resources, which puts constraints on the firm's decision to offer low-price products. To trade-off between all the attributes of software, namely, price, reliability, release time, testing time, sales, and profit. In this current work we have developed an optimization model including all these attributes together to compute the optimal launch time, the time till which testing will be carried out, and the development cost of the software under the aspiration level of reliability. Specifically, we first developed a sales model dependent on price and reliability. Second, we proposed a cost model incorporating market opportunity cost, release time, testing time, and sales function. Carrying these arguments further, the flow of the article is designed as follows: the modeling framework for modeling the sales and cost without and with post-release testing along with the list of assumptions has been given in Section 12.2. The validation and numerical investigation of the proposed framework is discussed in Section 12.3. The conclusion and future research have been given in Section 12.4, and the list of references are listed at the end of the chapter.

12.2 Modeling Framework

In this section, we will firstly describe the notation to be used during the modeling of the proposed framework.

12.2.1 Notations

$m(t)$	Expected number of faults removed by time t
a	Total faults lying in the software
T_{lc}	Software life cycle
b	Failure detection/correction rate
τ	Software release time
T	Testing stop time of the software
c_0	Setup cost
c_1	Market opportunity cost
c_2	Cost of debugging a fault by the testers before the release of the software
c_3	Cost of debugging a fault by the testers after the release when it is being reported by one of the users
c_4	Cost of debugging a fault by the testing team after release when the failure is detected by the testing team
c_5	Cost of debugging a fault reported by user after testing stop time
c_6	Cost of testing per unit time

12.2.2 Assumptions

Assumptions of the proposed model are as follows:

1. The fault removal follows the nonhomogeneous Poisson process.
2. Every failure is removed immediately and perfectly.
3. Software failure occurs due to remaining number of faults and the failure rate is equally affected by them.
4. Software consists of finite number of faults.
5. Life cycle of the software is finite.
6. Cost of patching is negligible.
7. Market opportunity cost is monotonically increasing, twice the continuously differentiable convex function of τ. We have used τ^2 as the functional form of market opportunity cost c_2 (Jiang et al., 2012).

On the basis of assumptions given in above in Section 12.2.2 the mean value function for total number of faults removed by time "t" is given by $m(t) = a(1 - e^{-bt}) = aF_1(t)$, where $F_1(t)$ is the cumulative fault removal rate before release. Also, for the proposed model we have used reliability of the software, which is defined as the ratio of number of faults detected at time "t" to the total number of faults present in the software (Lyu, 1996) and given by $R(t) = \dfrac{m(t)}{a}$.

12.2.3 Proposed Sales Model

It is impossible to debug all the faults lying inside the software. Also, long testing costs heavily and also delays the release. So, to remain in market it is important to determine the best time to release the software. This also takes care of the fact about handling the failures reported during operational phase and the time till which consumers are entertained so as to decide the testing stop time of the software. The optimization problem of determining the optimal price, release, and testing stop time can be formulated based on goals set by the management in terms of cost, reliability, resources, and so on. Although lot of research has been done on determining software release, due to changing environment it needs modification time to time. Users expect reliable software in no time, whereas firms aim at getting higher profit by minimizing software development cost. Keeping the above fact in mind, in this section we will present a software sales model dependent on price, release, and testing time and then develop the cost model for the software. Further, we will develop profit model with a constraint that is optimized to obtain optimal values of price, release, testing time, and profit from the sales. Description of the sales model is given below:

We assume that software sales (X) depend inversely upon price as well as on the unreliability of software and $\lambda = 1 - \dfrac{m(t)}{a}$ gives the unreliability of the software. On the basis of these two assumptions, software sales is given by

$$X = \theta e^{-\beta P} \lambda^{-\alpha} \tag{12.1}$$

Also $(\beta > 0)$ is the parameter reflecting the price influence, $\alpha > 0$ is the unreliability parameter, and factor θ represents the competitor and the other environmental influences. Now we develop the cost model based on the above sales function. Here software life cycle is divided into three phases: $[0, \tau]$, $[\tau, T]$, and $[T, T_{lc}]$. Description of the costs incurred in each phase is given below in Section 12.2.4.

12.2.4 Proposed Optimization Model

The cost model consists of different costs associated with different phases which are described below in Section 12.2.4.1 to 12.2.4.6.

12.2.4.1 Setup Cost

Software setup cost comprises the cost incurred during requirement, specification, design, and coding. All these phases come before the software release and ensure the customer needs and set the target for the testing team to develop the specific software. Setup cost usually depends on the development time and work effort spent during these phases. In the current study, we have assumed that setup cost is a function of software reliability parameter λ and is given by

$$E_0(T) = c_0 = S_1 + S_2\lambda^{-k} \tag{12.2}$$

where S_1 is the fixed (setup) cost and S_1, S_2, $k > 0$. From Equation (12.2), we see that with decreases in λ (or increase in reliability), setup cost increases. In other words, to get more reliable software, we have to invest more time and effort, which in turn increases the setup cost.

12.2.4.2 Market Opportunity Cost

It is the opportunity cost that a firm loses due to late release of the product. In simple words, it is the profit a firm could have gained by releasing its product earlier in the market by capturing the higher market share. In other way, it can be understood as the loss a firm bears between the predicted and actual release of product. We have taken a simple quadratic form to capture the impact of release on cost.

$$E_1(\tau) = c_1\tau^2 \tag{12.3}$$

12.2.4.3 Cost of Removal of Fault during Testing Phase before Release [0, τ]

The total number of bugs removed by the testing team before release is given by $m(\tau)$. Let c_2 be the cost of per unit removal of fault by the testing team; then, the total expected cost to remove $m(\tau)$ number of faults before software release is given by

$$E_2(\tau) = c_2 m(\tau) \tag{12.4}$$

12.2.4.4 Cost of Removal of Fault during Testing Phase after Release [τ, T]

This is an important phase that differentiates our model from the existing cost models. Here we have considered that testing continues post the release phase. In this phase, the faults that were left till the software release are detected and removed. This is to reduce the number of faults to give better experience to users during operational phase. During this phase, the testing team removes the faults on its own as well as the faults that were reported by users by experiencing failure during use. Both testers and users work for different aspects with different mind-sets during this phase: (1) testers look for the faults causing failure and (2) users look for completion of their own task. The rate of detecting the faults is also different. The detection rate of testers due to their knowledge and skill is more compared to that of the users. The rate of detection for users is $b.r$, where b is the detection rate of tester and r

is the ratio of fault detection rates of users to testers in $[\tau, T]$. On the basis of these arguments, the total number of faults removed in $[\tau, T]$ is given by

$$m(T - \tau) = a(1 - F_1(\tau))F_2(T - \tau) = ae^{-b\tau}\left(1 - e^{-(b+b')(T-\tau)}\right) \tag{12.5}$$

where $F_2(T-\tau)$ is the cumulative fault removal rate in the post-release phase and $b+b'$ is the increased rate of detection of faults due to an increase in the testing base, that is, combination of testers and users. Also, $a(1 - F_1(\tau))$ is the remaining number of faults after release time. Now the total number of faults detected by users out of $m(T-\tau)$ faults is

$$m'(T - \tau) = a(1 - F_1(\tau))F_2(T - \tau) = ae^{-b\tau}\left(1 - e^{-(b+b')(T-\tau)}\right) \cdot \frac{r}{r+1} \tag{12.6}$$

Cost incurred in this phase $[\tau, T]$ due to fault reported by user is given by

$$E_3(\tau, T) = c_3 \cdot m'(T - \tau)X^\delta \tag{12.7}$$

This cost is multiplied by the factor X^δ, as not all the customers come to the company for failure report and ask for fixing the fault removal during the warranty period. It can be understood by the following arguments: some of the customers will not find any failure during operational phase or some may find failure post testing stop time and so on. (Due to this reason, the value of δ will lie strictly between 0 and 1, $0 < \delta < 1$.) In the current work, we have considered $\delta = 0.5$ for numerical illustration purpose.

The total number of faults detected by testers out of $m(T - \tau)$ number of faults is

$$m''(T - \tau) = a(1 - F_1(\tau))F_2(T - \tau) = ae^{-b\tau}\left(1 - e^{-(b+b')(T-\tau)}\right) \cdot \left(1 - \frac{r}{r+1}\right) \tag{12.8}$$

The cost incurred for fixing a fault by tester in $[\tau, T]$ is

$$E_4(\tau, T) = c_4 \cdot m''(T - \tau) \tag{12.9}$$

12.2.4.5 Cost of Removal of Faults after Testing Stop Time T

The total number of faults detected by the users in this phase is given by

$$m(T_{lc} - T) = (a - m(T))F_3(T_{lc} - T) = ae^{-b\tau}e^{-(b+b')(T-\tau)}\left(1 - e^{-(T_{lc}-T)}\right) \tag{12.10}$$

where $(a - m(T))$ is the remaining number of faults after testing stop time. Hence, the cost of fixing these faults is

$$E_5(T) = C_5 \, m(T_{lc} - T) \cdot X^\delta \tag{12.11}$$

Multiplication of this cost by the factor X^δ has the same argument as given earlier.

12.2.4.6 Cost of Testing

The cost of testing is a linear function of time T. Hence, the total testing cost is given by

$$E_6(T) = c_6 \cdot T \tag{12.12}$$

12.2.4.7 Total Expected Software Cost

Combining all the above-mentioned costs together, we get the expected total software development cost function given by

$$E(\tau, T) = c_0 + c_1 \cdot \tau^2 + c_2 \cdot m(\tau) + c_3 \cdot m'(T - \tau) \cdot X^\delta + c_4 \cdot m''(T - \tau)$$
$$+ c_5 \cdot m(T_{lc} - T) \cdot X^\delta + c_6 \cdot T \tag{12.13}$$

12.2.4.8 Revenue from the Expected Sales is Given By

The total revenue obtained after sale of X units is

$$U(\tau, T, P) = P \cdot X \tag{12.14}$$

12.2.4.9 Expected Profit

Profit function for the above model is given by the difference between total revenue and costs, which is now a function of (1) sales price, (2) sales volume, and (3) development cost, and it is given by

$$\text{Profit} = Z(\tau, T, P) = U(\tau, T, P) - E(\tau, T) \tag{12.15}$$

We have also put the constraint on the expected sales and the reliability level to determine the optimal value for release, testing time, and price per unit of the software. Objective function of the proposed framework is given by

$$\text{Max Profit } Z(\tau, T, P)$$

$$\text{s.t.} \tag{12.16}$$

$$X \geq X_0, \lambda \leq \lambda_0$$

12.3 Numerical Illustration

To solve the proposed model to obtain optimal values, we have first estimated the parameters of the mean value function $m(t)$ on the data set given
by Wood (1996) using SPSS software. Data set contains a failure data set
for testing time of 20 weeks, and during this testing time, 100 faults were
detected. The parameter estimation result of the data set are $a = 130.201$,
$b = 0.083$. Combining all the values of cost and profit model, we get the functional form of the optimization model given as follows:

$$
Z(\tau, T, P) = P \cdot \theta e^{-\beta P} \lambda^{-\alpha} - \begin{bmatrix} S_1 + S_2 \lambda^{-k} + c_1 \cdot \leq \tau^2 + c_2 \cdot a \cdot (1 - e^{-b\tau}) \\[2mm] + c_3 \cdot ae^{-b\tau} \left(1 - e^{-(b+b')(T-\tau)}\right) \cdot \dfrac{r}{r+1} \cdot X^{\delta} \\[2mm] + c_4 \cdot ae^{-b\tau} \left(1 - e^{-(b+b')(T-\tau)}\right) \cdot \left(1 - \dfrac{r}{r+1}\right) + c_5 \\[2mm] \cdot ae^{-b\tau} e^{-(b+b')(T-\tau)} \left(1 - e^{-(T_{lc}-T)}\right) \cdot X^{\delta} + c_6 \cdot T \end{bmatrix} \dfrac{n!}{r!(n-r)!}
$$

subject to

$$X \geq X_0, \lambda \leq \lambda_0$$

(12.17)

For numerical illustration purpose, suppose $T_{lc} = 100$, $\beta = 0.006$, $\gamma = 0.06$,
$r = 0.5$, $s = 0.6$, $S_1 = 20,000$, $S_2 = 10,000$, $k = 3.5$, $c_1 = 5$, $c_2 = 30$, $c_3 = 100$, $c_4 = 30$,
$c_5 = 100$, and $c_6 = 50$, and the aspiration level for sales and reliability are
$X_0 = 25,000$ and $\lambda_0 = 0.2$. Note that $a = m(T_{lc})$. Substituting the given parameter values into Equation (12.17) and optimizing it using MAPLE software,
we get the results as follows: optimal price = 1668, optimal release time =
19.4 weeks, and optimal time to stop testing = after 71.7 weeks. The optimal
profit is 4.45×10^7 rupees. This means, to have a software product with reliability more than 80% and sales aspiration of 25,000, the firm should release
the software after 19.39 weeks of testing and stop testing the software after
71.7 weeks. One of the assumptions made during numerical illustration is
$\delta = 0.5$. To see the impact of this parameter on the optimal values, we have
carried out the sensitivity analysis on δ. The values of all the parameters
were kept constant and the values of δ were varied. The sensitivity analysis
results are given in Table 12.1.

   From the table, we see that on increasing the value of δ from 0.40 to 0.60,
a slight change is noted in the optimal values of price, the release time is
constant, but the testing stop time increases and the profit decreases. These
changes in the values of testing stop time and profit can be understood by the

TABLE 12.1

Sensitivity Analysis Results of Parameter δ

δ	Profit	Price	Release Time	Testing Stop Time
0.40	6.464427×10^7	1667.26	19.39	63.89
0.45	6.460224×10^7	1667.81	19.39	67.85
0.50	6.453019×10^7	1668.83	19.39	71.73
0.55	6.440794×10^7	1670.71	19.39	75.51
0.60	6.420051×10^7	1674.15	19.39	79.16

following argument. With an increase in the number of users reporting the failure during the operational phase, the cost of fixing it increases, thereby decreasing the profit of the firm. Also, due to more number of failures during the operational phase, the testing time has to be increased to debug the remaining number of faults, so that reliability can be increased.

12.4 Conclusion

Decisions related to software release, testing, pricing, and profit are some of the most important aspects of policy making in the software industry. Generally, firms have the tendency to delay the release of their software to ensure that the software has achieved a specified reliability level before release as it reduces the chance of software failure in field. But delay in release of software results in the loss of market opportunity. Pricing of the software also plays an important role to sustain in the competitive market situation. In our proposed optimization model, we have resolved this issue by proposing a profit model dependent on price, reliability, release, and testing time to obtain optimal values of profit, price release, and testing time. Specifically, we formulated a sales model dependent on price and testing and a cost model under different testing environments; formally analyzed the bug detection behavior under such a policy, accounting for periods when bugs are detected just by testers, as well as by both testers and users; and developed a profit model that can be used to determine the optimal profit by optimizing release and testing stop times and price. Therefore, if the price, release, and testing stop times are determined based on the proposed policy, project managers need not be concerned with an early-release decision, sales, and pricing of the software, because the risk will be more than offset by the benefits of post-release testing. This additional flexibility can help significantly accelerate the release time and help firms gain an advantageous position in a competitive marketplace. Furthermore, with existing release policies, new systems can be significantly "under tested," because project managers fear the consequences of delaying release. Our release policy, in contrast, allows firms to invest more on software testing, eventually leading to a more reliable system. Besides reducing the chances of failure in the field, a more reliable system helps the firm reap intangible benefits such as increased user satisfaction and customer loyalty. Future work may include other factors like resources and warranty in the sales model and their impact on the cost.

Acknowledgments

The research work presented in this chapter is supported by grants to the second author from University Grant Commission via RGNF-2012-13-SC-OR I-28203, Delhi, India.

References

Anand, A., Agarwal, M., Tamura, Y., & Yamada, S. (2017). Economic Impact of software patching and optimal release scheduling. *Quality and Reliability Engineering International*, 33(1), 149–157.

Arora A., Caulkins J.P., Telang R. (2006) Research note: Sell first, fix later: impact of patching on software quality, *Management Science*; Vol. 52, No. 3:465–471.

Cavusoglu H., Zhang J. (2008) Security patch management: Share the burden or share the damage?, *Management Science*; Vol. 54:657–670.

galorath.com/wp/software-project-failure-costs-billions-better-estimation-planning-can-help/ (accessed March 14, 2018).

HP. (2014) http://www.zdnet.com/hp-to-begin-charging-for-firmware-updates-and-service-packs-for-servers-7000026110, 2018 (accessed March 14, 2018).

Huang C.Y. (2005) Cost reliability optimal release policy for software reliability models incorporating improvements in testing efficiency, *Journal of System Software*; Vol. 77:139–155.

Huang C.Y., Lyu M.R. (2005) Optimal release time for software systems considering cost, testing-effort, and test efficiency, *IEEE Transactions on Reliability*; 54:583–591.

Jiang Z., Sarkar S., Jacob V.S. (2012) Post-release testing and software release policy for enterprise-level systems, *Information Systems Research*; Vol. 23, No. 3, Part 1 of 2, September:635–657.

Kapur P.K., Pham H., Gupta A., Jha P.C. (2011) *Software Reliability Assessment with OR Application*, Springer, Berlin.

Kapur P.K., Yamada S., Aggarwal A.G., Shrivastava A.K. (2013) Optimal price and release time of a software under warranty, *International Journal of Reliability, Quality and Safety Engineering*; Vol. 20, No. 3:14. DOI:10.1142/S0218539313400044.

Kapur, P.K., Shrivastava, A.K., Singh, O. (2017) When to release and stop testing of a software, *Journal of the Indian Society for Probability and Statistics*; Vol. 18, No. 1:19–37.

Kapur, P.K., Shrivastava, A.K. (2015) Release and testing stop time of a software: A new insight. In Reliability, Infocom Technologies and Optimization (ICRITO) (Trends and Future Directions), *2015 4th International Conference* on (pp. 1–7). IEEE.

Luo, C., Okamura, H., & Dohi, T. (2016). Optimal planning for open source software updates. Proceedings of the Institution of Mechanical Engineers, Part O: Journal of Risk and Reliability, 230(1), 44–53.

Lyu M.R. (1996) *Handbook of Software Reliability Engineering*, McGraw-Hill, Inc., Hightstown, NJ.

Musa J.D. (2004) Software reliability engineering: more reliable software, faster and cheaper.

Majumdar R., Shrivastava A.K., Kapur P.K., Khatri S.K. (2017) Release and testing stop time of a software using multi attribute utility theory, *Life Cycle Reliability and Safety Engineering*; Vol. 6, No. 1:47–55.

Okumoto K., Goel A.L. (1980) Optimum release time for software systems based on reliability and cost criteria, *Journal of System Software*; Vol. 1:315–318.

Pham H. (2003) Software reliability and cost models: Perspectives, comparison, and practice European, *Journal of Operational Research*; Vol. 149, No. 3:475–489.

Pham H. (2006) *System Software Reliability*, Springer-Verlag, London.

Sachdeva N., Kapur P.K., Singh O. (2018) Two-dimensional framework to optimize release time & warranty, In: *Quality IT and Business Operations: Modeling and Optimization*, pp. 383–404.

Ukimoto S., Dohi T. (2013) A software cost model with reliability constraint under two operational scenarios, *International Journal of Software Engineering and Its Applications*; Vol. 7, No. 3:236–239.

Wood A. (1996) Predicting software reliability, *IEEE Transactions on Computers*; Vol. 29:69–77.

www.gallop.net/blog/top-10-mega-software-failures-of-2014/ (accessed October 14, 2015).

www.alwaysagileconsulting.com/continuous-delivery-cost-of-delay/ (accessed October 14, 2015).

Xie M., Yang B. (2003) A study of the effect of imperfect debugging on software development cost, *IEEE Transactions On Software Engineering*; Vol. 29, No. 5:471–473.

Yamada S., Osaki S. (1987) Optimal software release policies with simultaneous cost and reliability requirements, *European Journal of Operational Research*; Vol. 31:46–51.

Yang B, Hu H, Jia L. (2008) A study of uncertainty in software cost and its impact on optimal software release time, *IEEE Transactions on Software Engineering*; Vol. 34:813–25.

Yun WY, Bai D.S. (1990) Optimum software release policy with random life cycle, *IEEE Transactions On Reliability*; Vol. 39, No. 2:167–170.

13

Software Reliability: A Quantitative Approach

Yashwant K. Malaiya

Colorado State University

CONTENTS

13.1   Introduction 205
13.2   The Software Life Cycle Phases 207
      13.2.1 Requirements.. 207
      13.2.2 Design... 208
      13.2.3 Coding 208
      13.2.4 Testing.. 208
      13.2.5 Operational Use 209
      13.2.6 Maintenance 209
13.3   Software Reliability Measures.. 210
13.4   Defect Detectability... 212
13.5   Factors that Impact Software Defect Density 212
13.6   Testing Approaches... 216
13.7   Input Domain Software Reliability Models 218
13.8   Software Reliability Growth Models... 219
      13.8.1 Planning Prior to Testing 222
13.9   Security Vulnerabilities in Internet Software 227
13.10 Evaluation of Multicomponent System Reliability 229
13.11 Optimal Reliability Allocation... 231
13.12 Tools for Software Testing and Reliability 234
13.13 Conclusions.. 234
References.. 235

13.1 Introduction

Software now controls everyday life of individuals involving commerce or social interactions, in addition to critical applications such as aviation or banking. This makes ensuring software reliability a major concern.

Software failures are known to cause enormous damage to the society, estimated to be in excess of a trillion dollar for 2017 by Tricentis [1]. During the past decades, very high reliability was expected mainly in the critical systems. Today software packages used every day also need to be highly reliable, because the enormous investment of the software vendor, as well as the user, is at stake. Studies have found that in software, reliability is considered to be the most important characteristic of a program by the customers.

It is virtually impossible to develop fault tree software unless the code needed is trivially small. All programs developed have defects regardless of how thorough the program development was. While tools have emerged that assist the software developers, program complexity has risen exponentially, sometimes doubling every year for some applications [2]. The programs must be tested for bugs and debugged until an acceptable reliability level is achieved. In any practical software system, the developer cannot assure that all the defects in it have been found and removed. However, the developer must ensure that the remaining bugs are few and do not have a catastrophic impact. Any software project must be released within the expected time to minimize loss of reputation and hence the market share. The developer must assess the risk associated with the residual bugs and needs to have a plan for achieving the target reliability level by the release date.

Quantitative methods are now coming in to use for obtaining and measuring reliability because of the emergence of validated and well-understood approaches. Data from industrial and experimental sources is now available that can be used to develop methods for achieving high reliability and validating them. The acceptable standards for software reliability levels have gradually risen in the past several decades.

This article takes the view that there exist several components and aspects of software reliability and they need to be integrated into a systematic framework. It presents the essential concepts and approaches needed for predicting, evaluating, and managing the reliability of software. The article first discusses the approaches taken during the successive phases of software development. Widely used measures of system software reliability and related metrics are next introduced. Detectability, a defect level metric, is introduced along with its implications. Factors that have been found to control software defect density are examined. Key ideas in test methodologies including the operational profile are presented. Nelson's model is presented in terms of the operational profile. Reliability growth in software is generally described using a software reliability growth model (SRGM). Here growth in reliability is illustrated and analyzed using actual test data. The use of such growth models allows estimation of the testing effort required to achieve the desired reliability goal. We also see how the reliability of a system can be evaluated if we know the failure rates of the components, and how reliability effort can be optimally allocated. Multi-version programming to achieve fault tolerance is also considered. Finally, the chapter presents the type of tools available for assisting in achieving and evaluating reliability.

The terms "failure" and "defect" are formally defined as follows [3]. The distinction is necessary to ensure a clear discussion.

Failure: a system behavior during execution that does not conform to the user requirements.

Defect: an error in software implementation that can cause a failure. It is also termed a fault or bug in literature.

A defect causes a failure when the code containing it is executed, and the resulting error propagates to the output. The probability of detecting a defect using a randomly chosen input is termed *testability* of a defect. Very low testability defects can be very hard to detect but can have a significant, perhaps even critical, impact on a highly reliable system.

As bugs are encountered and removed during testing, the reliability of the software improves. When the software project is released to the users, its reliability stays the same, as long as the same operating environment is used, and no patches are applied to modify the code. Bug fixes are often released in forms of patches updating the software, resulting in the next version of the code. For a program, its own past failure history is often a good indicator of its reliability, although data from other similar software systems may also be used for making projections [4].

13.2 The Software Life Cycle Phases

To achieve high reliability, reliability must be considered from the very beginning of software development. Different approaches are applicable during successive phases. The so-called waterfall model divides the software life cycle into these phases as discussed in the next paragraphs. However, for some development processes, it is common for developers to go to a previous phase, due to a need to make changes in the design or to respond to the required changes. It is much more cost-effective to find defects in an earlier phase because it can be significantly more expensive to address them later.

13.2.1 Requirements

In the requirements phase, the developing team interacts with the customer to understand and document the requirements the program needs. While in an ideal situation the requirements should specify the system unambiguously and completely, in actual practice, often corrective revisions during software development are needed. To identify missing or conflicting requirements, a review or inspection is conducted during this phase. It has been found that a significant number of bugs can be detected by this process. Any changes in the requirements in the later phases just before completion can increase the defect density.

13.2.2 Design

In the design phase, the overall architecture of the system is specified as an interconnection of small blocks termed *units*. The units should be well defined to be developed and tested independently. The design is reviewed to find errors.

13.2.3 Coding

In the coding phase, the actual code for each unit is first developed, generally in a higher level language such as Java, C, or C++. In the past, assembly level implementation was often required for high performance or for implementing input/output operations; however, with the emergence of highly optimizing compilers, the need to use assembly code has declined. The code developed is analyzed in team meetings to identify errors that can be caught visually.

13.2.4 Testing

This testing phase is the most significant phase when high reliability is needed and can take 30%–60% of the overall development effort. It is often divided into four phases:

1. *Unit testing*: In this phase, each unit is separately exercised using many test cases, and the bugs found are removed. As each unit is a relatively small piece of code, which can be tested independently, it can be tested much more exhaustively than a large program.
2. *Integration testing*: During this phase, the units are incrementally assembled, and at each step, partially assembled subsystems are tested. Testing subsystems permit the interaction among modules to be exercised. By incrementally incorporating the units within a subsystem, the unit responsible for a failure can be determined more easily.
3. *System testing*: In this phase, the completely integrated system is exercised. Debugging is continued until an appropriate exit criterion is

satisfied. During this phase, the emphasis is to find defects as quickly as possible. Researchers have pointed out that for highly reliable systems, the inputs should be drawn uniformly from the input partitions rather than using the frequencies that would be encountered during actual operation. However, when only limited testing can be afforded, perhaps due to an approaching deadline, actual operations should be simulated.

4. *Acceptance testing*: The objective of acceptance testing is to assess the system reliability and performance that would be encountered in the actual operational environment. This requires estimating the frequencies describing how the actual users would use the system. It is often termed alpha testing. Beta testing, which involves the use of the beta version released to some potential actual users, often follows next.

13.2.5 Operational Use

When the developer has determined that the suitable reliability criterion is satisfied, the software is released to the customers. In operational use, the bugs reported by the users are recorded so that they can be fixed when the next patch or bug fix is released.

13.2.6 Maintenance

After a release, the developer continues to maintain to code, until the retirement of the program in its current form. When major modifications or additions are made to an existing version, *regression testing* is done on the revised version to ensure that it still works and has not "regressed" to lower reliability because of potential incompatibility that may be injected. When a defect discovered is a security vulnerability, a patch for it needs to be developed as soon as possible. The time taken to develop and test a patch after the discovery of a vulnerability and the delayed application of an available patch can enhance the security risks. Support for an existing version of a software product needs to be offered until a new version has made a prior version obsolete. Eventually, maintenance is dropped when it is no longer cost-effective to the developing organization.

There can be some variation in the exact definitions of specific test phases and their exit criteria among different organizations. Some projects go through incremental refinements; for example, those using the extreme programming approach. Such projects may undergo several cycles of requirements–design–code–test phases.

The fraction of total defects introduced and found during a phase [5,6] will depend on the specific development process used. Table 13.1 gives

TABLE 13.1

Defects Introduced and Removed during the Successive Phases [7]

Phase	Defects (%)		
	Introduced	Found	Still Remaining
Requirements	10	5	5
Design	35	15	25
Coding	45	30	40
Unit testing	5	25	20
Integration testing	2	12	10
System testing	1	10	1

some representative fractions based on various studies. Majority of defects are encountered during design and coding. The fraction of defects found during the system test may be small; however, the phase can take a long time because the remaining defects require much more effort to be found. The testing phases can represent 30%–60% of the entire development effort.

13.3 Software Reliability Measures

In hardware systems, the reliability gradually declines because the possibility of a permanent failure increases. However, this is not applicable to software. During testing, the software reliability increases due to the removal of bugs and becomes constant when the defect removal is stopped. The most common software reliability measures include the following:

Availability: Using the classical reliability terminology, *availability* of a software system can be given as

$$A(t) = Pr\left\{\text{system is operating correctly at instant } (t)\right\} \qquad (13.1)$$

Transaction reliability: For software, the correctness of a single transaction is of major concern. A transaction-based reliability measure is often convenient to use.

$$R = Pr\left\{\text{a single transaction will not fail}\right\} \qquad (13.2)$$

Both these measures assume typical operation, that is, the input mix encountered obeys the *operational profile* as defined later.

Mean time to failure (MTTF): The mean time between two successive failures.

Failure intensity (λ): The mean number of failures per unit time is generally termed failure intensity. Note that failure intensity (per unit time) is inverse of MTTF. Thus, if the failure intensity is 0.005 per hour, MTTF is $1/0.005 = 200$ hours.

Since testing strategy is specifically designed to find the bugs at, failure intensity λ_t during testing is significantly higher, perhaps by one or two orders of magnitude, than the failure intensity λ_{op} encountered during normal operation. Test acceleration factor describes the relative effectiveness of the testing strategy as measured by the ratio λ_t/λ_{op}. Thus, if testing is 10 times more efficient for finding defects compared with the normal use, the test acceleration factor is 10. This factor is determined by the testing strategy and the type of application being tested.

Defect density: By convention, the software defect density is measured in terms of the number of defects per 1,000 source lines of code (KSLOC), not including the comments. Since many defects will not be found during testing, it cannot be measured directly, but it is possible to estimate it using the static and dynamic models discussed in the later sections. It can also be estimated using sampling approaches. The failure intensity depends on the rate at which the faults are encountered and thus can be assumed to be proportional to the defect density. The target defect density for a critical system or a high volume software can be equal to 0.1 defects/KLOC, or perhaps lower, whereas for other applications 0.5 defects/KLOC may be considered acceptable. Sometimes weights are assigned to defects depending on the severity of the failures they can cause. To keep analysis simple, here we assume that each defect has the same weight.

Test coverage metrics: Several tools are available that can evaluate how thoroughly a program has been exercised by a given test suite. The measurement is generally done by instrumenting the code during compilation. The two most common structural coverage metrics are as follows:

- *Statement coverage*: The fraction of all statements executed during testing. It is related to the metric termed block coverage.
- *Branch coverage*: The fraction of all branches that were taken during the code execution. It is related to the decision coverage metric.

The test coverage is correlated with the number of defects that will be encountered during testing [8]. 100% statement coverage is relatively easy to achieve; however, achieving 100% branch coverage can be hard for large programs. Often a predetermined branch coverage, say 85% or 90%, is used as an acceptance criterion. It becomes exponentially more difficult in terms of the number of test cases needed to achieve higher levels of branch coverage.

13.4 Defect Detectability

A key attribute of a defect is its detectability. The detectability of a specific defect can be defined as the probability that the defect will be tested by a randomly chosen test [9]. While some faults may be easily detected, some faults can be very hard to find. As testing progresses, the remaining faults become harder and harder to find. At any point in time during testing, the probability of detecting new faults depends on the *detectability profile*, which describes the number of remaining faults with the associated detectability values. During the later phases of testing, the larger number of remaining faults are likely to be the ones that are harder to find since easier faults would have already been found and removed [10]. Sometimes the faults found during limited testing are regarded to be representative of all the remaining faults. However, it is likely that they represent the easier-to-find faults and the faults not yet found are likely to be the ones that are harder to find.

The probability that a test will result in the detection of a fault is given by

$$P\{\text{a new fault is detected}\} = \sum_{i=1}^{n} N_i d_i \tag{13.3}$$

The detectability of a fault depends on how frequently the site of the fault is traversed and how easy it is for the resulting error to propagate to the output. Many programs contain code that is only executed when an error is encountered and a recovery is attempted. Faults in such sections of the code have a low probability of discovery with normal usage [11] or with random testing.

13.5 Factors that Impact Software Defect Density

The researchers in the field have attempted to identify the factors that impact the defect density and have evaluated their relative significance. The data available allows us to develop simple models for predicting their impact. Such a model can be used by a developer team to see how they can enhance the process and thus achieve higher reliability of their products. In addition, estimating the defect density can allow one to use a reliability growth model to estimate the testing effort needed. Here we consider the model by Malaiya and Denton [12], which was obtained using the data reported in the literature. Their model incorporates several factors and is given by

$$D = C \cdot F_{ph} F_{pt} F_m F_s F_{cc} F_{ru} \tag{13.4}$$

The following factors are used for estimation:

- The *phase factor* F_{ph}, which takes into account the *software test phase*.
- The *programming team factor* F_{pt}, which is based on the capabilities of the software development team.
- The *maturity factor* F_m, which depends on the maturity of the development process as given by the capability maturity model (CMM).
- The *structure factor* F_s, which depends on the structure of the software under development.
- The *code churn factor* F_{cc} which models the impact of requirement volatility.
- The *code reuse factor* F_{ru}, which takes in to account the influence of code reuse.

The factor C is a constant of proportionality representing the default defect density per KSLOC. The default value of each factor is 1. We propose the following preliminary submodels for each factor:

The phase factor F_{ph}: A simple model can be constructed using actual data reported by Musa et al. and Piwowarski et al. as shown in Table 13.2, where the multiplier values refer to the beginning of the phase specified. The numbers given indicate the variability in the factor; actual values may depend on the software development process. The numbers imply that the defect density is decreased by an overall factor of about 12 as a result of testing in various test phases. The beginning of the system test phase is taken as the default case with the value of 1.

The programming team factor F_{pt}: The coding and debugging capabilities of the individuals in a development team can vary significantly. Takahashi and Kamayachi [13] conducted a study correlating the number of defects with the programmer's experience in years. Their simple linear model can take into account programming experience of up to 7 years, each year reducing the number of defects by about 14%. Since formal studies on this subject are few, we can use Takahashi and Kamayachi's model as a guide to suggest the model in Table 13.3 based on the average skill level in a team. In addition

TABLE 13.2
Variability in the Phase Factor F_{ph}

Phase	Factor Value at the Beginning of the Phase
Unit testing	4
Subsystem testing	2.5
System testing	1 (default)
Operation	0.35

TABLE 13.3
Variability in the Programming Team Factor F_{pt}

Programmers' Skill Level	Factor Value
Strong	0.4
Average	1 (default)
Weak	2.5

to programming experience, personal discipline and understanding of the problem domain can also influence defect density. Table 13.3 assumes that worst programmer correspond to the most inexperienced programmers, as studied in Ref. [13].

The capability maturity factor F_m: The Software Engineering Institute (SEI) CMM measures the maturity of the software process at an organization, which can range from ad hoc (level 1) to an optimizing one (level 5) where data is continually collected and used to refine the process. Table 13.4 uses the data collected by Jones and Keene as well as that reported in a Motorola study. Most organizations are believed to be at level 1; however, the table uses level II as the default level, since a level I organization is not likely to be using software reliability engineering.

The software structure factor F_s: This factor considers various aspects of the software structure including program complexity, language type (the fractions of code in assembly and high-level languages), and module size distribution. The assembly language code is harder to write and thus naturally will have a higher defect density. The impact of complexity, as measured by several complexity metrics, has been extensively investigated by researchers. Most complexity metrics have been found to be strongly correlated to software size as measured in the number of lines of code (LOCs). Since the defect density is being considered here, which is the number of defects divided by software size, software size has already been discounted.

The language dependence can be given by a simple model for F_s by considering a, the fraction of the code that is in assembly:

$$F_s = 1 + 0.4a$$

TABLE 13.4
Variability in the Process Maturity Factor F_m

CMM Level	Factor Value
Level 1 (Initial)	1.5
Level 2 (Repeatable)	1 (default)
Level 3 (Defined)	0.4
Level 4 (Managed)	0.1
Level 5 (Optimizing)	0.05

This assumes that assembly language code has 40% more bugs.

Any software project is composed of a collection of modules of various sizes. A module may be a function, a class, a file, and so on. The size of a module has some impact on the defect density. Very large modules may be hard to comprehend. When module sizes are very small, they can have a higher fraction of bugs related to the interaction among them. It has been found that module sizes tend to be asymmetrically distributed with a larger number of small modules. In such a case, module size distribution can impact defect density [14]. Additional research is needed to construct a model for this factor.

The code churn factor F_{cc}**:** This factor causes changes in the code because of changes in the requirements (also termed requirement volatility) [15,16]. It can be shown that this is a function of the fraction of code f_c that is changed and the time t_c when the change occurs as follows:

$$F_{cc}(t_c) = (1 - f_c) + f_c \, e^{\beta 1 t_c}$$

where $\beta 1$ is a parameter. The impact is higher when the change occurs later.

The code reuse factor F_{ru}**:** It is common for software projects to use some preexisting code that has already undergone prior testing and thus has a lower defect density [17]. If the defect densities of the reused and new codes are d_r and d_n, respectively, and if u is the fraction of code that is inherited, then it can be shown that factor is given by

$$F_{ru} = 1 - u + u \frac{d_r}{d_n}$$

The resulting defect density is lower if more of the code is reused.

Calibrating the defect density model: The multiplicative model given in Equation (13.4) can provide an early estimate of the defect densities. The factor C needs to be estimated using past data from preferably the same organization for similar projects. Other factors not considered here will be taken into account implicitly by calibrating the overall model. Since the beginning of the subsystem test phase is taken as the default in the overall model, an examination of the Musa's data sets suggests that the constant of proportionality C can range from about 6–20 defects per KSLOC. Some indeterminacy is inherent in such models. It can be accounted for by using an upper and a lower estimate and using both of them to make predictions.

Example 1

The value of the constant C has been found to range between 12 and 16 for an organization. A team in the organization is working on a new project. The team has similar capabilities as other past teams. About 80% of the code is in a high-level language, and the rest 20% is in assembly language. Half of the project reuses the previous code, which has one-fourth the defect density compared with the new code. Other factors are assumed to be what is typical for the organization. The software size is about 20,000 LOC. What is the expected number of total defects at the beginning of the integration test phase?

We note that assembly code increases the defect density by a factor of $(1 + 0.4 \times 0.2) = 1.08$, and reuse decreases the defect density by a factor of $(0.5 + 0.5 \times 0.25) = 0.625$. The integration test phase implies a factor of 2.5 relative to the beginning of the system test phase. From the model given by Equation (13.4), we estimate that the defect density at the beginning of the subsystem test phase can range between $12 \times 2.5 \times 1 \times 1 \times 1.08 \times 1 \times 1 \times 0.625 = 20.25/\text{KSLOC}$ and $16 \times 2.5 \times 1 \times 1 \times 1.08 \times 1 \times 1 \times 0.625 = 27/\text{KSLOC}$. Thus, the total number of defects can range from 405 to 540.

13.6 Testing Approaches

Testing a program involves applying a number of inputs and observing the responses. When the response is different from the expected, the code has some (one or more) defects. During debugging, the objective of software testing is to find faults as quickly as possible and remove them to increase the reliability as fast as possible. However, during certification, the object is to estimate the reliability during actual deployment; thus, the fault encountering rate should be representative of the actual operation. The testing approaches can be divided into these two classes:

A. *Black-box (or functional) testing*: This assumes that nothing about the implementation of the software is known. The test generation is done by only considering the functional input/output description of the software. This is the often most common form of testing.

B. *White-box (or structural) testing*: In this case, the information about the actual implementation is used, for example, the existence of specific blocks or branches. The testing then can target specific structural elements of the code.

A combination of the two approaches will often yield the best results. Black-box testing only needs a functional description of the software. However, specific information about actual implementation will allow testers to better select elements of the structure that need to be exercised. When the inputs are selected randomly, the approach is termed *random testing*. When *partition testing* approach is used, the input space is divided into partitions defined using the ranges of the input variables/structures. The test cases are then chosen to ensure that each partition is exercised thoroughly. Random testing and partition testing are orthogonal approaches and can be combined.

Partitions can be exercised randomly for the usual cases and deterministically for boundary cases.

Many faults have high *testability*, that is, they are easily detected with only limited testing. Some faults that are triggered only under rarely occurring input combination will have very low testability. At the beginning of testing, many faults are found quickly. They are faults that have high testability and are easily detected and removed. Since the easily detectable faults have already been removed, the remaining faults are found in the later phases of fault removal. Finding the hard-to-test faults can take considerable effort. When a very low defect density product is needed, the testers need to use careful and systematic approaches.

A test coverage measure can measure using the thoroughness of testing, as discussed in Section 13.3. Branch coverage is a more thorough measure than statement coverage, since complete branch coverage guarantees complete statement coverage, but not vice versa. A branch coverage of say 85% may be used as a minimum criterion. A stricter measure (like p-use coverage) or a combination of measures (such as those provided by the test coverage tools) should be useful for very high-reliability programs.

Test inputs must be chosen and drawn in accordance with the *operational profile* to be able to estimate the reliability in the field. A *profile* is the set of disjoint operations that a software performs, with their probabilities of occurrence. The *operational profile* is based on the frequencies that occur in actual operation. When a software is used in very different user settings, the operational profile for each setting may be different. For obtaining an operational profile, the input space needs to be divided into sufficiently small partitions and then estimating the probabilities of an input being drawn from a partition. A partition with a high associated probability may need to be further divided into smaller sub-partitions to provide enough resolution.

Example 2

This example is based on an example by Musa [18]. A system responds differently to a call depending on the type of incoming transaction. Using prior experience, the following transaction types (TTs) and maintenance operation types (MOTs) are identified and their probabilities have been determined as follows:

A.	TTA	0.74
B.	TTB	0.15
C.	MOTC	0.10
D.	MOTD	0.009
E.	MOTE	0.0005
F.	MOTF	0.0005
G.	Failure recovery	0.000001
Total for all operation types		1.0

It is noted that TTA occurs 74% of the time, and thus the profile does not have enough resolution. It was found that the transaction type TTA can be further divided depending on the subtypes, given as follows:

A1	TTA1	0.18
A2	TTA2	0.17
A3	TTA3	0.17
A4	TTA4	0.12
A5	TTA5	0.10
Total for all subtypes of TTA		0.74

Thus, the final operational partitions are {A1, A2, A3, A4, A5, B, C, D, E, F, G}. These and their probabilities form the operational profile given by {(A1, 0.18), (A2, 0.17) ... (G, 0.000001)}. When acceptance testing is performed, the test operations would be chosen such that a transaction type TTA1 occurs 18% of the time, TTA2 occurs 17% of the time, and so on. While failure recovery occurs only rarely, it is likely to be a critical operation and thus needs at least some test effort.

If the testing objective is to estimate the failure rate, as would be the case during acceptance testing, testing should be done according to the operation profile. Also, operational profile-based testing would be efficient if the testing time is very limited. However, if the project needs to achieve sufficiently high reliability, testing needs to more uniform across operation types. Often code that handles failures and exceptional cases will contain faults that will be hard to detect. For such defects, special tests need to be used so that the code containing them is entered and exercised [19].

13.7 Input Domain Software Reliability Models

An early input domain model was proposed by Nelson [20]. If there are I distinct possible inputs, and a failure occurs for I_e of them, then the reliability is given by

$$R = 1 - \frac{I_e}{I} \tag{13.5}$$

If the software input space can be partitioned into partitions $j = 1, 2,..., k$, and the probability of an input being chosen from partition j is p_j as given by the operational profile, then Equation (13.5) can be refined as

$$R = 1 - \sum_{j=1}^{k} p_j \frac{I_{ej}}{I_j} \tag{13.6}$$

where *Ij* is the number of inputs applicable for partition *j* and I_{ej} is the number of inputs for which an error is encountered. This makes a simplifying assumption that failures for different partitions are statistically dependent. In actual practice, at least some code would be shared, and thus Equation (13.6) should be viewed as an approximation. It can be used to estimate the reliability when the operational profile is known along with the failure rate I_{ej}/I_j for a partition.

The Nelson's model assumes that the software is not changing. When the software is being tested and debugged, the reliability grows as bugs are gradually removed. The relationship between the debugging effort and the resulting reliability growth is discussed in the next section.

13.8 Software Reliability Growth Models

In a software development project, the cost needed to ensure a suitable reliability level can often be 25%–40%, although for very high reliability needs, it has been reported to be as much as 60%. A software needs to be released by a specific target date because of customer requirements and competitive marketing. The manager must ensure that the software has achieved the desired quality level before it can be released. Even after extensive testing, any further testing will always detect result in finding more bugs. The manager must determine the quality level at which the cost of a future encounter with residual bugs is less than the cost of additional testing. The manager thus must determine the amount of testing needed so that the minimum acceptable reliability level will be achieved. Careful planning in terms of time and effort needed requires the use of modeling the defect finding and removal process. A number of such models, termed SRGMs, have been proposed.

An SRGM assumes that reliability will grow with testing time *t*, which can be measured in terms of the CPU execution time used, or the number of man-hours or days. The time can also be measured in terms of the number of transactions encountered. Here we consider the exponential model, which can be thought to represent several models, including the models proposed by Jelinski and Muranda (1971), Shooman (1971), Goel and Okumoto (1979), and Musa (1975–1980). Here it is derived using basic considerations, even though many other models are obtained by noting the trend in the actual data collected.

The SRGMs are generally described using one of these two functions of testing time—*failure intensity* $\lambda(t)$ or *total expected faults* (often termed the *mean value function*) detected by time *t*, given by $\mu(t)$. The failure intensity is the rate of change in the number of a cumulative number of faults detected by time *t*:

$$\lambda(t) = \frac{d}{dt}\mu(t) \qquad (13.7)$$

Let us denote the number of yet undetected defects at time t to be $N(t)$. Here we assume that debugging is perfect, implying that a defect is always successfully removed when it is encountered.

The exponential model assumes that at any time, the rate of finding (and removing) bugs is proportional to the number of undetected defects still present in the software. That can be stated as follows:

$$-\frac{dN(t)}{dt} = \beta_1 N(t) \tag{13.8}$$

where β_1 is a constant of proportionality. The parameter β_1 can be shown to be given by

$$\beta_1 = \frac{K}{\left(S \cdot Q \cdot \dfrac{1}{r}\right)} \tag{13.9}$$

where S is the software size (number of source instructions), Q is the number of object instructions per source instruction as a result of compiling a high language code into machine language for the target instruction set architecture, and r is the instruction execution rate for the machine instructions of the computer being used. The term K called *fault exposure ratio* has been found to be in the range of 1×10^{-7} to 10×10^{-7} when t is measured in seconds of CPU execution time. It is notable that K does not depend on the software size but can depend on the testing strategy. As a first approximation, we regard K to be constant.

The differential equation (13.8) can be solved to yield

$$N(t) = N(0)e^{-\beta_1 t} \tag{13.10}$$

where $N(0)$ is the initial number of faults before testing started. Hence the total number of expected faults detected by time t is given by

$$\mu(t) = N(0) - N(t)$$
$$= N(0)(1 - e^{-\beta_1 t}) \tag{13.11}$$

In the software reliability engineering literature, it is generally written in the form:

$$\mu(t) = \beta_0(1 - e^{-\beta_1 t}) \tag{13.12}$$

and thus β_0 is the number of faults that would eventually be found, which is equal to $N(0)$ when we assume perfect debugging, that is, no new defects are generated during debugging.

Using Equation (13.7), we can obtain an expression for failure intensity given in terms of the two parameters β_0 and β_1:

$$\lambda(t) = \beta_0 \beta_1 e^{-\beta_1 t} \tag{13.13}$$

One significant advantage of the exponential model is that it has a simple explanation and both parameters β_0 and β_1 have a clear interpretation and thus can be estimated even prior to testing. The variations of the exponential model include the hyperexponential model, proposed by Ohba, Yamada, and Lapri, which assumes that different sections of the software are separately governed by an exponential model, with the parameter values specific for each section.

Many other SRGMs have been proposed and used [21]. Several studies have compared the models for both fit and predictive capability using the data from different projects. It has been found that the exponential compares well with other models, allthough, a couple of models have been found to outperform the exponential model. We will here look at one of them, the logarithmic model, proposed by Musa and Okumoto (termed logarithmic Poisson model by them), which has been found to have a better predictive capability than the exponential model.

The logarithmic model assumes that the fault exposure ratio K varies during testing [22]. The logarithmic model is also a finite-time model, implying that after a finite (although perhaps long) time, there will be no more faults to be found. The logarithmic model models a logarithmic dependence on time and can be given by

$$\mu(t) = \beta_0 \ln(1 + \beta_1 t) \tag{13.14}$$

or in terms of the failure intensity

$$\lambda(t) = \frac{\beta_0 \beta_1}{1 + \beta_1 t} \tag{13.15}$$

The preceding equations apply as long as $\mu(t) < N(0)$. The condition is practically always satisfied since testing has to be terminated while a few bugs are still present, since finding additional bugs gets exponentially harder.

The variation in K, as assumed by the logarithmic model has been observed in actual practice. For higher defect densities, the value of K declines with testing, as defects get harder to find. However, K starts rising at low defect densities. This is explained by the fact that practical testing tends to be directed toward functionality not yet adequately exercised, rather than being random. This impacts the behavior at low defect densities.

For the logarithmic model, the two parameters β_0 and β_1 do not have a straightforward interpretation. An interpretation is provided by Malaiya and Denton [12] in terms of the variation in fault exposure ratio. An approach for estimating the logarithmic model parameters β_0^L, β_1^L has been proposed by relating them to the exponential model parameters.

The exponential model has been shown to underestimate the number of defects that will be detected in a future interval, that is, it tends to have a negative bias. The logarithmic model also exhibits a negative bias, although somewhat smaller. Only the Littlewood–Verral Bayesian model exhibits a positive bias among the major models. This model also has good predictive capabilities; however, it is not popular because it is relatively more complex, and the parameter values do not have a physical interpretation.

An SRGM is needed in two management situations. A manager can use it before testing begins to estimate the testing needed to achieve the desired reliability. That needs an estimation of the parameters using static attributes like the software size. When testing and debugging is being done, actual test data can be collected and used to estimate the parameter values and to make projections.

13.8.1 Planning prior to Testing

Project managers need to come up with a preliminary strategy for testing early so that testers can be hired. For the exponential model, the two values of the two parameters β_0 and β_1 can be designed using defect density model and Equation (13.9) respectively. They can then be used to estimate the testing effort that would be required to reach the target reliability attribute, which can be defect density, failure intensity, or MTTF.

Example 3

For a software development project, the defect density prior to testing has been estimated to be 25 defects/KLOC. The software size is 10,000 lines written in C. The code expansion ratio Q for C code for a target CPU is about 2.5, and thus the compiled object code size will be $10,000 \times 2.5 = 25,000$. The testing uses a CPU with MIPS rating of 70 (i.e., 70 million machine instructions per second). The fault exposure ratio K is expected to be 4×10^{-7}. Obtain the testing time required such that a defect density of 2.5 defects/KLOC will be achieved.

Using the exponential model, the value of the first parameter can be estimated as

$$\beta_0 = N(0) = 25 \times 10 = 250 \text{ defects,}$$

The second parameter can be estimated this way:

$$\beta_1 = \frac{K}{\left(S \cdot Q \cdot \dfrac{1}{r} \right)} = \frac{4.0 \times 10^{-7}}{10,000 \times 2.5 \times \dfrac{1}{70 \times 10^6}}$$

$$= 11.2 \times 10^{-4} \text{ per second}$$

If t_1 is the testing time required to reduce the defect density to 2.5/KLOC, then Equation (13.10) can be used to obtain its value

$$\frac{N(t_1)}{N(0)} = \frac{2.5 \times 10}{25 \times 10} = \exp(-11.2 \times 10^{-4} \cdot t_1)$$

giving us

$$t_1 = \frac{-\ln(0.1)}{11.2 \times 10^{-4}} = 2056 \text{ seconds CPU time}$$

The failure intensity at the end of the testing period will be

$$\lambda(t_1) = 250 \times 11.2 \times 10^{-4} e^{-11.2 \times 10^{-4} t_1}$$

$$= 0.028 \text{ failures/second}$$

The example provides a simple illustration. It has been shown that the value of K depends on the testing strategy used. The above example uses CPU seconds as the unit of time and hence the effort. Often the effort is specified in terms of the number of person-hours. The person-hours can be converted into the equivalent CPU execution time by using by an appropriate factor, which would be obtained using past data. A convenient way to estimate β_1 would be to note that Equation (13.9) suggests that $\beta_1 \times I$ should be constant provided only the code size is different. For example, if for a previous project with 10 KLOC source code, the final value for β_1 was 2×10^{-3} per second. Then for a new 20 KLOC project, β_1 can be estimated as $2 \times 10^{-3}/2 = 1.00 \times 10^{-3}$ per second.

USING PARTIAL TEST DATA

During testing, the defect finding rate can be recorded as testing progresses. By using an SRGM, a manager can predict the additional testing time needed to achieve the desired reliability level, or determine the testing effort needed if there is a target release date. The major steps for analyzing the partial data are as follows.

1. *Collection and conditioning of data*: The failure intensity data points always include a lot of short-term variations. The data needs to be smoothed to extract and preserve the long-term trend. Data can be *grouped* by dividing the test duration into a number of intervals and then computing the average failure intensity in each interval.
2. *Model selection and fitting*: The best model that fits the partial data many not be the best choice since the testing effectiveness changes as testing progresses. The preferred way to select a model might be to rely on the past experience with other projects using the same process. The exponential has the advantage that the parameters are amenable to estimation using static data. The logarithmic models have been found to have good

predictive capability across various projects. Early test data is very noisy, and the parameter values may not stabilize until a significant amount of data is available. The parameter values can be determined using either least square or maximum likelihood approaches.

3. Using the parameter values, an analysis can be performed to decide how much more testing is needed. The fitted model can project the testing effort needed to achieve the target failure intensity or defect density. Recalibrating a model that does not conform to the data can significantly improve the accuracy of the projection. A model that describes the process well may not improve significantly by recalibration.

Example 4

The data in this example was collected by Musa [3] for an industrial project with 21,700 instructions. Testing was continued for about 25 hours and 136 defects were found. In Table 13.5, the time is specified in seconds when the next set of 5 defects were found. The failure intensity is computed for each set of 10 defects. This results in some smoothing of data.

Fitting all the data points to the exponential model, the parameter values are obtained as

$$\beta_o = 150.62 \text{ and } \beta_1 = 3.4 \times 10^{-5} \text{ per second}$$

The exact values will depend on the fitting approach used. Let the target failure intensity be one failure per hour, that is, 2.0×10^{-4} failures per second. An estimate of the stopping time t_f is then given by

$$2.0 \times 10^{-4} = 150.92 \times 3.4 \times 10^{-5} e^{-3.4 \times 10^{-4} t_f} \tag{13.16}$$

which gives the total testing time needed as $t_f = 95,376$ seconds., that is, 26.49 hours. This means an additional testing of about 2 hours. In Figure 13.1, the smooth curve shows the fitted model.

The variation of the parameter values of the exponential model has been studied. Results show that the value of β_0 estimated early during testing and debugging is generally lower than the final value. The final value of β_1 tends to be lower than the initial value. Thus, as testing progresses, the estimated value of β_0 rises and β_1 declines, and the product $\beta_0\beta_1$ remains stable. Thus, the early estimates are likely to be fewer but with higher testing efficiency. In Equation (13.16), the actual value of t_f is likely to be higher. Thus, the answer of 15.69 hours should be taken to represent a low estimate for the test time needed.

In many cases, it is desirable to compute interval estimates of the testing time and the final failure intensity. There exist statistical methods that can be used for calculating the interval estimates [19,20]. Sometimes the testing needs to continue until a failure rate is achieved with a specific confidence level, say 90%. Graphical and analytical approaches for obtaining such stopping points have been developed [21].

TABLE 13.5

Software Failure Data

Defects Found	Time (seconds)	Failure Intensity (per second)
5	342	
10	571	0.015974441
15	968	0.007067138
20	1,986	0.004694836
25	3,098	0.003264773
30	5,049	0.004492363
35	5,324	0.007513148
40	6,380	0.004310345
45	7,644	0.002696145
50	10,089	0.002995806
55	10,982	0.004048583
60	12,559	0.00399361
65	13,486	0.003079766
70	15,806	0.002517623
75	17,458	0.002666667
80	19,556	0.001499475
85	24,127	0.002025522
90	24,493	0.000789141
95	36,799	0.00062162
100	40,580	0.001819174
105	42,296	0.001164009
110	49,171	0.000945269
115	52,875	0.001371366
120	56,463	0.001022913
125	62,651	0.000685871
130	71,043	0.000498728
135	82,702	0.000340155
136	88,682	

The growth models assume that the testing strategy is uniform throughout testing. In actual software testing, the test approach is altered from time to time. Each approach is initially very efficient for detecting a specific class of faults, causing a spike in failure intensity as a result of the switch. This requires a smoothing approach to minimize the impact of these spikes during fitting. The models assume that the software under test is stable during testing; however, often it changes because of additions or changes in an evolving software. To minimize the error due to changes, early data points can be dropped from the calculation, using only the more recent data points. If reliability data for each component added or deleted can be separately assessed, the approaches discussed in the section on multicomponent systems may be used.

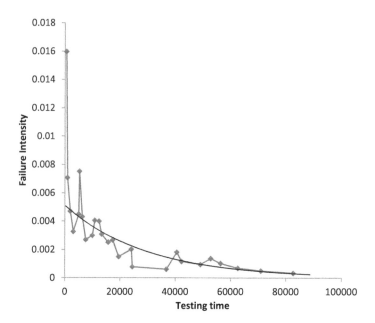

FIGURE 13.1
SRGM fitting.

USING TEST COVERAGE TO EVALUATE RELIABILITY GROWTH

Several software tools are available that can evaluate coverage of statements, branches, and other structural and data flow elements (such as P-uses). In large programs, it may not be feasible to achieve complete (100%) coverage, still, a higher test coverage implies the code has been exercised more thoroughly.

The software test coverage is directly related to the residual defect density and hence the reliability [8]. It can be shown that μ, the total number of defects found, is linearly related to the test coverage measures at higher test coverage. For example, if the branch coverage C_B is being evaluated, it will be found that for low values of C_B, μ remains close to zero. However, for higher values of C_B (after a bend in the curve termed a *knee*), μ starts rising linearly. The model is given by

$$\mu = -a + b \cdot C_B, \quad C_B > \text{knee} \tag{13.17}$$

In the preceding equation, the parameters a and b will be influenced by the software size and the defect density at the beginning. It is notable that the variations in test effectiveness will not influence the relationship since the test coverage directly evaluates the thoroughness of testing of a program. When the system is expected to have a high reliability, a stricter metric like branch coverage should be used rather than statement coverage.

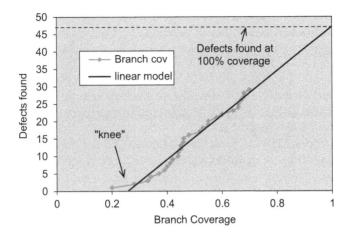

FIGURE 13.2
Coverage-based modeling.

Figure 13.2 gives the actual data from a European Space Agency project. When testing was stopped, 29 defects were found and 70% branch coverage was obtained. According to the model, if the testing had continued until 100% branch coverage was obtained, 47 total defects would have been found. Thus, about 18 residual defects were present when testing was stopped. According to the test data, only one defect was detected when branch coverage was about 25%; thus, the knee of the curve can be considered to be at a branch coverage of 0.25.

13.9 Security Vulnerabilities in Internet Software

Software systems such as operating systems, servers, and browsers can present major security challenges because systems are now generally connected using a network or the Internet and the number of security breaches is increasing. The software defects in such programs that permit an unauthorized action are termed *vulnerabilities*. Reported vulnerabilities are tracked by the CERT and other databases. Modern society, from individuals to commercial and national organizations depend on the Internet connectivity. That makes the exploitation of vulnerabilities very attractive to criminals or adversaries with the technical expertise. Vulnerabilities and their exploits can now be bought and sold in the international arena.

Each software system goes through several phases: the release, increasing use, and eventual stability followed by a decline in the market share when it is replaced by a new version or a new system. The three phases in the cumulative vulnerabilities plot for a specific version of the software can often be clearly observed. When a product is first released, there is a slow rise. As the product develops a market share, it starts attracting the attention of both white hat and black hat vulnerability finders. This creates a steady pace of vulnerability discovery reports. When an alternative program takes away the market share, the vulnerability finding rate declines in an older program. Figure 13.3 gives a plot of vulnerabilities reported for Windows 98 during January 1999–August 2002.

A model proposed by Alhazmi and Malaiya [22] for the cumulative number of vulnerabilities, y, against calendar time *t*, can be given by

$$y = \frac{B}{BCe^{-ABt} + 1} \tag{13.18}$$

In the equation, *A*, *B*, and *C* are the parameters to be determined from data. The parameter *B* represents the total number of vulnerabilities present that will be eventually found. The data for several common operating systems fit the model as determined by the goodness of fit values.

Just as the overall defect density is a major reliability metric, the *vulnerability density* in a large software system is an important measure of risk. The known vulnerability density can be computed using the complete data about the reported vulnerabilities. Known vulnerability density V_{KD} is the reported number of vulnerabilities in the software divided by the software size. Thus,

$$V_{KD} = \frac{V_K}{S} \tag{13.19}$$

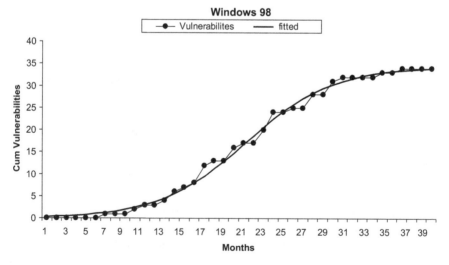

FIGURE 13.3
Vulnerability discovery in Windows 98.

TABLE 13.6
Vulnerability Densities and Ratios for Windows NT 4.0 and Windows 98

System	MSLOC	Known Defects (1,000s)	D_{KD} (per KLOC)	Known Vulnerabilities	V_{KD} (per KLOC)	Ratio (V_{KD}/D_{KD})
Windows NT 4.0	16	10	0.625	206	0.0129	2.06%
Windows 98	18	10	0.556	84	0.0047	0.84%

where V_K is the reported number of vulnerabilities and S is the size of software in the number of LOC. Table 13.6 gives the values based on data from several sources [22]. It gives the known defect density D_{KD}, V_{KD}, and the ratios of the two.

In Table 13.6, we see that the source code size is 16 and 18 MSLOC (million source lines of code) for Windows NT 4.0 and Windows 98, respectively, which is approximately the same. The reported defect densities at release, 0.625 and 0.556, are also similar for the two OSs. The known vulnerabilities in Table 13.6 are as of the period 1999–2008 (no vulnerabilities were reported after that). For Windows 98, the vulnerability density was 0.0047, whereas for Windows NT 4.0 it is 0.0129, nearly three times. The higher vulnerability density for Windows NT 4.0 is likely due to two factors. First, since it is a server OS, it contains more code that handles external accesses. Second, attacking servers would generally be much more rewarding, and thus it must have attracted more attention by the vulnerability finders causing detection of more vulnerabilities.

The last column in the table gives the ratios of known vulnerabilities to known defects. For the two systems, the ratios are 1.06% and 0.84%. A few researchers had presumed that of the total defects, vulnerabilities can be 1% or 5%. The values given in Table 13.6 show that 1%–2% may be closer to reality.

13.10 Evaluation of Multicomponent System Reliability

A large software system is always implemented using a number of modules [24], which may be separately developed by different teams. The individual modules are likely to have been developed and tested differently, some possibly reused from a previous version, causing a variation in defect densities and failure rates. Here we discuss methods for computing the system level failure rate and the reliability if we know the reliabilities attributes of the individual modules. In a system where only one thread executes at a time, one specific module is under execution at a time. Some modules are invoked more often, and some may remain under execution for a longer time. If f_i is the fraction of the time module i being executed, then the average system failure rate is

$$\lambda_{sys} = \sum_{i=1}^{n} f_i \lambda_i \qquad (13.20)$$

where λ_i represents the failure rate of module i.

Let the execution time of a single transaction be T. Let us assume that module i is invoked e_i times the transaction time T, and each time execution takes for d_i time, then its execution frequency is

$$f_i = \frac{e_i \cdot d_i}{T} \qquad (13.21)$$

The system reliability R_{sys} can be specified as the probability that no failures will occur during a single transaction. From the reliability theory, the system reliability is given by

$$R_{sys} = \exp(-\lambda_{sys} \cdot T)$$

Using the above mentioned equations, we can write the expression as

$$R_{sys} = \exp\left(-\sum_{i=1}^{n} e_i d_i \lambda_i\right)$$

The single execution reliability of module i is R_i is $\exp(-d_i \lambda_i)$. Thus, we have

$$R_{sys} = \prod_{i=1}^{n} (R_i)^{e_i} \qquad (13.22)$$

Multiple-version programming: In the critical domains, like defense or avionics, sometimes redundant versions of the program are employed. To reduce the probability of multiple numbers of them failing at the same time, each version is developed independently by a separate team. The implementation can use triplication and voting on the result to achieve fault tolerance. The system is operating correctly as long as two of the versions are working correctly. The scheme assumes the voting mechanism to be defect free. If the failures in the three versions are independent as ideally expected, the reliability can be improved by a large factor. However, it has been shown that that there is a significant probability of correlated failures that must be considered.

Here is a simple analysis. In a three-version system, let the probability of all three versions failing for the same input be q_3. Also, let the probability that any two versions will fail at the same time be q_2. As three distinct pairs are possible among the three versions (three choose two), the probability P_{sys} of the system failure for a transaction is

$$P_{sys} = q_3 + 3q_2 \tag{13.23}$$

If the failures are statistically independent, and if the probability of one version failing is p, Equation (13.23) can be written as

$$P_{sys} = p^3 + 3(1-p)p^2 \tag{13.24}$$

Experiments have demonstrated that in reality, the failures have some statistical correlation. The values of q_3 and q_2 then need to be experimentally determined for system reliability evaluation.

Example 5

The potential improvement in system reliability in an n-version system can be illustrated using the experimental data collected by Knight and Leveson, as has been pointed out by Hatton [25]. For the experimental data, the probability of a version failing for one transaction was 0.0004. If there is no correlation, then a three-version system would have an overall failure probability given by P_{sys}:

$$P_{sys} = (0.0004)^2 + 3(1-0.0004)(0.0004)^2$$

$$= 4.8 \times 10^{-7}$$

This represents an improvement by a factor of $0.0004/4.8 \times 10^{-7} = 833.3$. However, a significant correlation was observed in the experiments, causing the probability of all three modules failing at the same time to be $q_3 = 2.5 \times 10^{-7}$ and the probability of any two modules failing to be $q_2 = 2.5 \times 10^{-6}$. Using these values, we get

$$P_{sys} = 2.5 \times 10^{-7} + 3 \times 2.5 \times 10^{-6} = 7.75 \times 10^{-6}$$

Thus, the realistic improvement factor achieved is $0.0004/7.75 \times 10^{-6} = 51.6$.

Hatton argues that the best software testing approaches have been found to reduce defect density only by a factor of 10. Thus, an improvement factor 52 is not achievable by additional testing; it can only be obtained by using a redundant n-version software implementation.

13.11 Optimal Reliability Allocation

A large program is always developed as a set of separate units that are eventually integrated to form the complete system. The separate modules will have different sizes and defect densities, especially if some of them are reused. That gives rise to the problem of allocating the test effort to different modules in a way that minimizes the total testing cost, while still obtaining the target reliability level. This is an application of the reliability allocation problem that arises in mechanical and electrical systems also. The problem is easier for a software system because the reliability can be continuously improved by using additional testing, and the impact of testing can be modeled by the SRGMs.

Assuming that the exponential SRGM is applicable, the failure rate can be specified as a function of testing time d_i for a module i as discussed. The failure rate can be rewritten as

$$\lambda_i(d) = \lambda_{0i} \exp(-\beta_{1i} d_i)$$

where the SRGM parameters applicable to the module i are $\lambda_{0i} = \beta_{0i} \beta_{1i}$ and β_{1i}. Thus, the test cost, measured in terms of the testing time needed, d_i, can be expressed as a function of the module's failure intensity:

$$d(\lambda_i) = \frac{1}{\beta_i} \ln\left(\frac{\lambda_{0i}}{\lambda_i}\right)$$

To obtain the overall system failure rate, we assume that the block i is under execution for a fraction f_i of the time ($\Sigma f_i = 1$). The reliability allocation problem can be formally stated as an optimization problem

$$\text{Minimize the total cost } C = \sum_{i=1}^{n} \frac{1}{\beta_i} \ln\left(\frac{\lambda_{0i}}{\lambda_i}\right) \tag{13.25}$$

$$\text{subject to } \lambda_{ST} \leq \sum_{i=1}^{n} f_i \lambda_i \tag{13.26}$$

where the desired overall failure intensity is given by λ_{ST}. The optimization problem given by Equations (13.25) and (13.26) can be solved by using the Lagrange multiplier approach [26] to obtain a closed form solution. The optimal failure rates and the testing times needed are obtained as follows:

$$\lambda_1 = \frac{\dfrac{\lambda_{ST}}{f_1}}{\displaystyle\sum_{i=1}^{n} \dfrac{\beta_{11}}{\beta_{1i}}} \quad \lambda_2 = \frac{\beta_{11} f_1}{\beta_{12} f_2} \lambda_1 \quad \cdots \quad \lambda_n = \frac{\beta_{11} f_1}{\beta_{1n} f_n} \lambda_1 \tag{13.27}$$

The analysis results in the optimal values of d_1 and d_i, $i \neq 1$ as given by

$$d_1 = \frac{1}{\beta_{11}} \ln \left(\frac{\lambda_{10} f_1 \sum_{i=1}^{n} \frac{\beta_{11}}{\beta_{1i}}}{\lambda_{ST}} \right) \quad and \quad d_i = \frac{1}{\beta_{1i}} \ln \left(\frac{\lambda_{i0} \beta_{1i} f_i}{\lambda_1 \beta_{1i} f_1} \right) \quad (13.28)$$

Note that the testing time for a block must be nonnegative. The value of d_i is positive if $\lambda_i \leq \lambda_{i0}$.

The parameter β_{1i} of the exponential model as given by Equation (13.9) is inversely proportional to the software size as discussed earlier. As a first assumption, the value of f_i can be assumed to be proportional to the module size. Also, we note that the values of λ_i and λ_{i0} depend on the initial defect densities, but not on size. Thus, Equation (13.27) implies that the optimal values of the failure rates $\lambda_1, \dots \lambda_n$, after testing and debugging, will be equal. If the initial defect densities are same for all the blocks, then the optimal test times for each module would be proportional to its size. In many applications, some of the blocks are more critical. That can be modeled by assigning additional weights resulting in longer testing times for them.

Example 6

A software system uses five functional modules M1 to M5. To illustrate optimization, we construct this example assuming sizes 1, 1, 3, 00, and 20 KLOC, respectively, for the five modules, with the initial defect densities of 5, 10, 10, 15, and 20 defects per KLOC. The assumptions result in the parameter values are given in the top three rows, which are the inputs to the optimization problem. The optimal solution obtained using the aforementioned equations are given in the two bottom rows. The same result would be obtained if a numerical optimization is done (e.g., using Excel Solver). Let us now compute the optimal distribution of the additional testing time required for each module to achieve the overall system failure rate equal to 0.15 while minimizing the total test cost. Here the time units can be hours of testing time or hours of CPU time used for testing.

Module	M_1	M_2	M_3	M_4	M_5
β_i	5×10^{-3}	5×10^{-3}	1.67×10^{-3}	2.5×10^{-4}	2.5×10^{-4}
λ_{i0}	0.2	0.4	0.4	0.4	0.8
x_i	0.022	0.022	0.067	0.444	0.444
Optimal λ_i	0.15	0.15	0.15	0.15	0.15
Optimal d_i	57.49	196.1	588.5	3923	6696

The results show that the final values of λ_i for all the modules are equal, even though they begin with different initial values. This requires a larger fraction of the test effort allocated to blocks that are larger (compare d_2 and d_3) or with higher defect densities (compare d_1 and d_2). The cost in terms of total additional testing effort needed is 11,461.41 hours.

13.12 Tools for Software Testing and Reliability

Software reliability is an engineering discipline. Thus, it requires collection, analysis of data, and the evaluation of proposed designs. Many tools are now becoming available that can automate several of the tasks such as testing, evaluation of testing, assessing potential reliability growth and failure rates. Some of the representative types of tools are mentioned here, with an example to illustrate. Some of the tools are in public domain. Installing and learning to use a tool can require a significant amount of time; thus, a tool should be selected after a careful comparison of the applicable tools available. Tables are available in the literature that compare the features or performance of different tools [27,28].

- Automatic test generations: JCONTRACT
- Graphical User Interface (GUI) testing: AscentialTest
- Memory testing: Valgrind
- Defect tracking: Bugzilla
- Test coverage evaluation: JCover
- Reliability growth modeling: Statistical Modeling and Estimation of Reliability Functions for Systems (SMERFS)
- Defect density estimation: ROBUST
- Coverage-based reliability modeling: ROBUST
- Markov reliability evaluation: Symbolic Hierarchical Automated Reliability and Performance Evaluator (SHARPE)
- Fault tree analysis: Eclipse Modeling Framework based Fault-Tree Analysis (EMFTA)

13.13 Conclusions

This chapter presents an overview of the basic concepts and approaches for the quantitative software reliability field. It considers the software life cycle and discusses the factors that impact the software defect density. Use of an SRGM is presented and illustrated using actual data. It also discusses a model for the discovery of security vulnerabilities in Internet-related software. Reliability evaluation of a system using the failure rates of the components is presented, and an approach is given for optimal allocation of reliability. The article also mentions the type of tools that can be used to assist in achieving and evaluating reliability.

References

1. W. Platz, *Software Fail Watch*, 5th ed., Tricentis, Vienna, 2018.
2. B. St. Clair, Growing complexity drives need for emerging DO-178C standard, *COTS The Journal of Military Electronics and Computing*, 2009.
3. J. D. Musa, *Software Reliability Engineering*, McGraw-Hill, New York, 1999.
4. Y. K. Malaiya and P. Srimani (Ed.), *Software Reliability Models*, IEEE Computer Society Press, Los Alamitos, CA, USA, 1990.
5. D. Carleton, R. E. Park and W. A. Florac, Practical software measurement, Tech. Report, SRI, CMU/SEI-97-HB–003.
6. P. Piwowarski, M. Ohba and J. Caruso, Coverage measurement experience during function test, *Proceedings, International Conference on Software Engineering*, 1993, pp. 287–301.
7. Y. K. Malaiya, Software reliability management, In: S. Lee, (Ed.), *Encyclopedia of Library and Information Sciences*, 3rd ed., Vol. 1: Taylor & Francis Group, Milton Park, 2010, pp. 4901–4912
8. Y. K. Malaiya, N. Li, J. Bieman, R. Karcich and B. Skibbe, The relation between test coverage and reliability, *Proceedings, IEEE-CS International Symposium on Software Reliability Engineering*, Nov. 1994, pp. 186–195.
9. Y. K. Malaiya, A. von Mayrhauser and P. Srimani, An examination of fault exposure ratio, *IEEE Transactions on Software Engineering*, 19(11), pp. 1087–1094, 1993.
10. E. N. Adams, Optimizing preventive service of software products, *IBM Journal of Research and Development*, 28(1), pp. 2–14, 1984.
11. H. Hecht and P. Crane, Rare conditions and their effect on software failures, *Proceedings, IEEE Reliability and Maintainability Symposium*, Los Angeles, CA, Jan. 1994.
12. Y. K. Malaiya and J. Denton, What do the software reliability growth model parameters represent, *Proceedings, IEEE-CS International Symposium on Software Reliability Engineering ISSRE*, Nov. 1997, pp. 124–135.
13. M. Takahashi and Y. Kamayachi, An empirical study of a model for program error prediction, *Proceedings, International Conference on Software Engineering*, Aug. 1995, pp. 330–336.
14. Y. K. Malaiya and J. Denton, Module size distribution and defect density, *Proceedings, International Symposium on Software Reliability Engineering*, Oct. 2000, pp. 62–71.
15. N. Nagappan and T. Ball, Use of relative code churn measures to predict system defect density. *Proceedings, 27th International conference on Software engineering (ICSE '05)*, ACM, 2005, pp. 284–292.
16. Y. K. Malaiya and J. Denton, Requirements volatility and defect density, *Proceedings, 10th International Symposium on Software Reliability Engineering (Cat. No.PR00443)*, Boca Raton, FL, 1999, pp. 285–294.
17. T. J. Ostrand and E. J. Weyuker, The distribution of faults in a large industrial software system. *Proceedings, ACM SIGSOFT International Symposium on Software Testing and Analysis (ISSTA '02)*, 2002, pp. 55–64.
18. J. Musa, More reliable, faster, cheaper testing through software reliability engineering, Tutorial Notes, ISSRE '97, 1997, pp. 1–88.

19. H. Yin, Z. Lebne-Dengel and Y. K. Malaiya, Automatic test generation using checkpoint encoding and antirandom testing, *International Symposium on Software Reliability Engineering*, 1997, pp. 84–95.

20. T. A. Thayer, M. Lipow and E. C. Nelson, *Software Reliability*, North-Holland Publishing, TRW Series of Software Technology, Amsterdam, 1978.

21. M. R. Lyu (Ed.), *Handbook of Software Reliability Engineering*, McGraw-Hill, Hightstown, NJ, USA, 1996, pp. 71–117.

22. N. Li and Y. K. Malaiya, Fault exposure ratio: Estimation and applications, *Proceedings, IEEE-CS International Symposium on Software Reliability Engineering*, Nov. 1993, pp. 372–381.

24. P.B. Lakey and A. M. Neufelder, System and software reliability assurance notebook, Rome Lab, FSC-RELI, 1997.

25. L. Hatton, N-version design versus one good design, *IEEE Software*, 14(6), pp. 71–76, 1997.

26. Y. K. Malaiya, *"Reliability Allocation" Encyclopedia of Statistics in Quality and Reliability*, John Wiley & Sons, Hoboken, NJ, 2008.

27. Comparison of GUI testing tools, (updated time to time) available at https://en.wikipedia.org/wiki/Comparison_of_GUI_testing_tools.

28. S. Wang and J. Offutt, Comparison of unit-level automated test generation tools, in *Fifth Workshop on Mutation Analysis*, 2009, pp. 2–10.

14

Modeling and Analysis of Fault Dependency in Software Reliability Growth Modeling

V. B. Singh

University of Delhi

CONTENTS

14.1 Introduction 237
14.2 Fault Dependency Concept 238
14.3 Software Fault Detection and Correction Process 239
    14.3.1 Modeling of Fault Detection and Removal Processes 240
    14.3.2 Power Function of Testing Time–Based Fault Detection
        and Removal Processes 241
14.4 Fault Dependency–Based Software Reliability Growth Models 242
    14.4.1 Error Dependency Model 242
    14.4.2 Considering Time Lag in Removal of Leading Faults and
        Fault Dependency 243
    14.4.3 Fault Dependency Models by Relaxing Time Lag in
        Leading Faults 244
    14.4.4 Power Function of Testing Time–Based Fault Dependency
        Models by Relaxing Time Lag in Leading Faults 246
    14.4.5 Power Function–Based Fault Dependency Models by
        Considering Time Lag in Leading Faults 247
14.5 Conclusion 250
References 250

14.1 Introduction

During software testing, efforts are made to uncover the defects lying dormant in the software. The dormant faults are of different nature and complexities. It is necessary to remove these defect prior to release of the software. In some cases, fault correction is not performed immediately once a failure is detected. A fault has a life cycle consisting of reporting of faults on fault repositories, its verification to see that it is really a fault, root cause analysis, and creating a patch to fix the fault. Once a fault is removed/fixed, it is closed. This fixed fault can also be reopened by some users if it is found to be incorrectly fixed. During fault identification, once the attributes of faults are identified, fault-fix time prediction models have been used to predict the time that can be taken to fix the fault. There is a time lag between fault detection and its removal. The removal of faults depends upon many factors such as the team who writes the code to fix the faults and how skilled they are. The nature of the fault is an important factor that affects the debugging time required during the software fault removal. Schneidewind [1–3] presented the concept of fault detection and correction as a two-stage process through their modeling. The author proposed that there is a constant time delay lag between detection and correction of the fault. The idea of the constant delay time lag has been revisited and a time-dependent time lag function has been proposed by Xie and Zhao [4]. The authors observed that as the testing progresses, the faults that are detected during later stage of the fault detection/removal process requires more efforts for its removal. In this chapter, we try to understand the fault detection and correction process by a mathematical modeling. We determine the time lag between fault detection and correction process. By using the fault dependency concept and debugging time lag, various software reliability growth models have been presented in the chapter. These lag functions have been derived from different fault detection and correction–based software reliability growth models.

14.2 Fault Dependency Concept

During software development, efforts have been made for early fault prediction and removal through various techniques. A software consists of faults of different nature; some may be severe from a developer's point of view and others severe from a user's point of view. During software testing, we also found that the faults are not only severe, but also have the nature of being leading/independent and dependent. Leading faults are independent in nature, which means its removal does not depend upon any other faults. We can directly remove those faults. Dependent faults are those whose removal depends upon the removal of faults on which it is dependent. To model dependent and independent faults in software reliability growth modeling, first a model was proposed, as described in Ref. [5]. The proposed model in [5] has been revised and has amalgamated time lag functions in the modeling of dependent faults [6]. Of late, by considering debugging time lag functions and fault dependency concept, models have been proposed by researchers, namely, Huang et al. [7], Kapur et al. [8,9], Singh et al [10], and Chatterjee et al. [11]. The following example explains the concept of independent and dependent faults in a software.

A software consists of many functions and these functions consist of a set of programs written in any computer language. This is also called source code. This source code has sequential statements, conditional statements, and control statements. In sequential statements, a large number of variables, operators are used. In many cases, these statements are linked together and the correct output of one statement depends upon the correct output of its preceding statements. Then, we say that these statements have a dependency, that is, the correct execution of one statement is prevented by the other statement.

In sequential statements, fault dependency exists due to some variable used in different statements. The following table explains the fault dependency in a sequential statement of code.

Line	Statement	Correct Program	Faulty Program	Remarks
10	S-1	a = m+1	a = m−1	Wrong operator
20	S-2	g = r%10	g = r%15	Wrong operand
30	S-3	s = 200%a	s = 200/a	Wrong operator
40	S-4	z1 = a%g	z1 = g%a	Reverse order

From the above mentioned set of statements, we found that wrong operator, wrong operand, and the reverse of the operand have been used. Left-side statements are correct and right-side statements are faulty. Fault lying in statement line number 30 can only be fixed once we fixed the fault in line number 10. Similarly, the statement in line number 40 can only give the correct output once we fix the faults in statements line numbers 10 and 20. In the similar line, we can also explain the concept of fault dependency using conditional and repetitive statements.

14.3 Software Fault Detection and Correction Process

On the bug tracking system, bugs are reported by different users sitting across different geographical locations. Once the bugs are detected, it takes time to analyze and hence there is a delay. During software testing efforts are also made to remove all the faults prior to releasing the software to customers. Once we detect faults, it is not corrected immediately. In the literature, several authors addressed the issue of delayed fault correction time. An approach for modeling the fault correction process by using a constant delayed fault detection process has been proposed in the literature and many enhancements have been made following the model.

In the following sections, we are discussing how to model the time delay between detection and correction of a fault.

14.3.1 Modeling of Fault Detection and Removal Processes

Once a fault is reported on a bug tracking system and it is analyzed, then real cause is identified and after that it is removed/corrected/fixed. It is important to note that this delay may not be a constant but time-dependent. In many cases, it may follow a distribution function too. In an actual software environment, detected fault is not immediately removed, but it lags the fault detection process by a delay factor $\phi(t)$. The delay effect factor $\phi(t)$ is the debugging time lag function assumed to be a time-dependent function that measures the expected delay in correcting a detected fault at any time.

Now it is clear to us that detection and correction process exist in the software debugging process. We can construct the existing software reliability growth models as a delayed model. We have explained how existing models can be constructed as a delayed model in the following sections. We have also interpreted some existing software reliability growth model as the delayed fault detection model.

Let the mean value function of the expected number of faults detected and corrected in time interval [0,t] be defined as

$$m_f(t) = a\left(1 - \exp(-bt)\right)$$

and

$$m_r(t) = m_f\left(t - \phi(t)\right)$$

or

$$m_r(t) = a\left(1 - \exp\left(-b\left(t - \phi(t)\right)\right)\right) \tag{14.1}$$

where $m_r(t)$ is the cumulative faults removed up to time t. $m_f(t)$ is the cumulative faults detected up to time t, and a is the potential number of faults lying dormant in the software. b is the fault detection rate per remaining faults.

On the basis of Equation (14.1), we will interpret the following well-known established models as delayed models by considering the concept of debugging time lag as follows in [9,10]:

Case 1: if we consider $\phi(t) = 0$, the model in Equation (14.1) will be reduced to $a\left(1 - \exp(-b(t))\right)$, that is, GO model [12].

Case 2: If we consider $\phi(t) = \left[\dfrac{1}{b}\log(1 + bt)\right]$ in Equation (14.1), the model will be reduced to Yamada model [13], that is, $m_r(t) = a\left(1 - (1 + bt)\exp(-bt)\right)$.

Case 3: If we take $\phi(t) = \left[\dfrac{1}{b}\log\left(1 + bt + \dfrac{b^2t^2}{2}\right)\right]$ in Equation (14.1), then we will have the following model [14], that is, $m_r(t) = a\left[1 - \left(1 + bt + \dfrac{b^2t^2}{2}\right)\exp(-bt)\right]$.

Case 4: If we take $\phi(t) = \left[\dfrac{1}{b} \log\left(\dfrac{(1+\beta)\exp(-bt)}{1+\beta\exp(-bt)} \right) \right]$ in Equation (14.1), then we

will have the following model [15], that is, $m_r(t) = a\left(\dfrac{1-\exp(-bt)}{1+\beta\exp(-bt)} \right)$, where β

is a constant.

14.3.2 Power Function of Testing Time–Based Fault Detection and Removal Processes

By considering the fact that number of faults detected and removed is a function of the number of instructions executed and as a power function of testing time as follows in [10,16,17], we have

$$m_r(t) = m_f\left(t - \phi(t)\right)$$

and

$$m_r(t) = a\left(1 - \exp\left(-\frac{b}{k+1}\left(t - \phi(t)\right)^{(k+1)} \right) \right) \tag{14.2}$$

where k is a constant.

Case 1: If we take $\phi(t) = 0$ in Equation (14.2), we will have a model presented in [16], that is, $m_r(t) = a\left(1 - \exp\left(-\dfrac{b}{k+1}t^{(k+1)} \right) \right)$.

Case 2: If we take $\phi(t) = t - \left[t^{k+1} - \dfrac{k+1}{b}\log\left(1 + \dfrac{bt^{k+1}}{k+1} \right) \right]^{\frac{1}{k+1}}$ in Equation (14.2),

we will have a model presented in [16], that is, $m_r(t) = a\left(1 - \left(1 + b\dfrac{t^{k+1}}{k+1} \right) \right.$

$\left. \exp\left(-b\dfrac{t^{k+1}}{k+1} \right) \right)$.

Case 3: If we take $\phi(t) = t - \left[t^{k+1} - \dfrac{k+1}{b}\log\left(1 + \dfrac{bt^{k+1}}{k+1} + \dfrac{b^2}{2}\left(\dfrac{t^{k+1}}{k+1} \right)^2 \right) \right]^{\frac{1}{k+1}}$

in Equation (14.2), it will be reduced to that presented in [16], that is,

$m_r(t) = a\left(1 - \left(1 + b\dfrac{t^{k+1}}{k+1} + \dfrac{b^2}{2}\left(\dfrac{t^{k+1}}{k+1} \right)^2 \right)\exp\left(-b\dfrac{t^{k+1}}{k+1} \right) \right)$.

Case 4: If we take $\phi(t) = t - \left[-\dfrac{1}{b}\log\left(\dfrac{(1+\beta)\exp\left(-\dfrac{bt^{k+1}}{k+1}\right)}{1+\beta\exp\left(-\dfrac{bt^{k+1}}{k+1}\right)} \right)^{k+1} \right]^{\frac{1}{k+1}}$ in Equation

(14.2), we will have a model presented in [16], that is, $m_r(t) = a\dfrac{\left(1-\exp\left(-\dfrac{bt^{k+1}}{k+1}\right)\right)}{\left(1+\beta\exp\left(-\dfrac{bt^{k+1}}{k+1}\right)\right)}.$

14.4 Fault Dependency–Based Software Reliability Growth Models

Software reliability growth models have been widely used to assess the reliability growth of software products. In industries and research and development organizations, these reliability growth models have been applied widely in managerial decision making. An effort has always been made to develop the software that is close to the software development paradigm and internal structure of the software product. Fault dependency-based software reliability growth models have been presented in the literature. The developed models have considered assumptions that are very near to software development and, especially, testing phase. Some common assumptions have been used by different researchers for developing models by considering fault dependency and time lag between detection and correction of a fault. Some models that represent the fault dependency concept are discussed in Sections 14.4.1–14.4.5.

14.4.1 Error Dependency Model

First, Kapur and Younes [5] proposed a software reliability growth model that considers the fault/error dependency concept. The model is as follows:

The potential fault/error in the software is a, which is the sum of leading and dependent errors, that is, $a = a_1 + a_2$, where a_1 and a_2 are the number of leading and dependent errors, respectively.

The errors are removed from the software in time interval $[t, t + \Delta t]$ and let it be denoted by $m(t)$.

The removal of leading and dependent errors is also assumed to follow a nonhomogeneous Poisson process (NHPP). Thus, $m(t)$ can be written as the superposition of two NHPP$_s$:

$$m(t) = m_1(t) + m_2(t) \tag{14.3}$$

The dormant errors lying in the software were detected and fixed. Let $m_1(t)$ be the content of leading errors and $m_2(t)$ be the content of dependent errors. These errors are removed in the time interval $[0,t]$.

With the assumption that the detection of leading errors depends upon the number of leading errors remaining in the software, then, we can write the following differential equation:

$$\frac{dm_1(t)}{dt} = b \times [a_1 - m_1(t)]$$

Solving this differential equation under the boundary condition, that is, at $t = 0$, $m_1(0) = 0$, we get $m_1(t) = a_1(1 - \exp(-bt))$, where b is the leading error removal rate per remaining leading error. The given equation models the leading fault removal phenomenon.

With the assumption that dependent fault removal depends upon the remaining dependent fault removal and to the ratio of leading fault removed to the total number of faults with a time lag T, we have the following differential equation:

$$\frac{dm_2(t)}{dt} = c \times [a_2 - m_2(t)] \times \frac{m_1(t-T)}{a}$$

In this model, c is the rate of removal for dependent error/fault. The removal of dependent error is lag with time T with the removal of leading error. To get the solution of the above equation, T is taken to be zero/negligible. We have the following model for total fault removal using Equation (14.3).

$$m(t) = a\left(1 - p\exp[-bt] - (1-p)\exp\left[\frac{pc}{b}(1 - \exp[-bt]) - ptc\right]\right) \quad (14.4)$$

During dependent error removal modeling, the author has taken into consideration the "ratio of leading errors removed to the total error at time t," but later in Ref. [6], the authors have modified this assumption and taken the "ratio of leading errors removed at time t to the total number of leading errors." The model is given as follows:

$$m(t) = a\left(1 - p\exp[-bt] - (1-p)\exp\left[\frac{c}{b}(1 - \exp[-bt]) - tc\right]\right) \quad (14.5)$$

14.4.2 Considering Time Lag in Removal of Leading Faults and Fault Dependency [7]

Huang et al. [7] incorporated the assumption given in [5] in their modeling and proposed the following models. In reference [7], it is assumed that leading faults also have the time lag in their removal process.

For leading faults:

$$\frac{dm_1(t)}{dt} = b \times [a_1 - m_1(t)], \text{ with the initial condition, that is, at } t = 0, m_1(0) = 0,$$

we have $m_1(t) = a_1(1 - \exp[-bt])$.

By considering the following equation for dependent faults,

$$\frac{dm_2(t)}{dt} = c \times [a_2 - m_2(t)] \times \frac{m_1(t - \phi(t))}{a} \tag{14.6}$$

We are presenting the following models proposed in [7] by using different debugging time lag functions $\phi(t) = \left[\frac{1}{b}\log(1 + bt)\right]$ and $\phi(t) = \left[\frac{1}{b}\log\left(\frac{(1+\beta)\exp(-bt)}{1+\beta\exp(-bt)}\right)\right]$. Using Equation (14.3) and the solution of Equation (14.6) at initial condition $t = 0$, $m_2(0) = 0$, we have the following mean value functions for total fault removal:

$$m(t) = a\left(1 - p(1 + bt)\exp[-bt] - (1 - p)\right.$$

$$\left.\exp\left[\frac{2pc}{b}(1 - \exp[-bt]) - tpc(1 + \exp[-bt])\right]\right)$$

and

$$m(t) = a\left(1 - p\frac{(1+\beta)\exp(-bt)}{1+\beta\exp(-bt)} - (1 - p)\exp[-pct]\left(\frac{1+\beta}{1+\beta\exp[-bt]}\right)^{\frac{pc(1+\beta)}{b\beta}}\right)$$

This model also considers the debugging time lag during leading fault removal.

14.4.3 Fault Dependency Models by Relaxing Time Lag in Leading Faults [9]

By considering that leading faults may immediately be removed, no debugging time lag is required for the removal of leading faults. Next, we have the models presented in [9].

We have the following equation for dependent faults with the assumption that "dependent faults removal depends upon the remaining dependent faults and to the ratio of leading faults removed to the total leading faults." Also assuming that leading faults have no debugging time lags unlike in [7], we have the following equations:

For leading faults:

$$\frac{dm_1(t)}{dt} = b \times [a_1 - m_1(t)] \tag{14.7}$$

For dependent faults:

$$\frac{dm_2(t)}{dt} = c \times [a_2 - m_2(t)] \times \frac{m_1(t - \phi(t))}{a_1} \tag{14.8}$$

Solving these equations at boundary conditions and by considering $\phi(t) = 0$ and using Equation (14.3), we have the following model:

$$m(t) = a\left(1 - p\exp[-bt] - (1-p)\exp\left[\frac{c}{b}(1 - \exp[-bt]) - tc\right]\right)$$

Various debugging time lag functions have been derived (Section 14.3.1) and are being used for modeling dependent faults.

Model 1: If $\phi(t) = \left[\frac{1}{b}\log(1 + bt)\right]$, then by solving Equations (14.7) and (14.8) at boundary conditions and by using Equation (14.3), we will get the following model:

$$m(t) = a\left(1 - p\exp[-bt] - (1-p)\exp\left[\frac{2c}{b}(1 - \exp[-bt]) - tc(1 + \exp[-bt])\right]\right)$$

Model 2: If $\phi(t) = \left[\frac{1}{b}\log\left(1 + bt + \frac{b^2t^2}{2}\right)\right]$, then by solving Equations (14.7) and (14.8) at boundary conditions and by using Equation (14.3), we will get the following model:

$$m(t) = a\left(1 - p\exp[-bt] - (1-p)\exp\left[\begin{array}{c}\frac{3c}{b}(1 - (1 + bt)\exp[-bt]) \\ -tc\left(1 - \left(1 - \frac{bt}{2}\right)\exp[-bt]\right)\end{array}\right]\right)$$

Model 3: If $\phi(t) = \left[\frac{1}{b}\log\left(\frac{(1+\beta)\exp(-bt)}{1 + \beta\exp(-bt)}\right)\right]$, then by solving Equations (14.7) and (14.8) at boundary conditions and by using Equation (14.3) we will get the following model:

$$m(t) = a\left(1 - p\exp(-bt) - (1-p)\exp[-ct]\left(\frac{1+\beta}{1 + \beta\exp[-bt]}\right)^{\frac{c(1+\beta)}{b\beta}}\right)$$

14.4.4 Power Function of Testing Time–Based Fault Dependency Models by Relaxing Time Lag in Leading Faults [9]

Considering the fault detection process to be dependent on residual fault content and on power function of testing time. In this section, various models have been presented for leading and dependent faults: Software reliability growth models have been developed incorporating fault dependency with various debugging time lag functions using a power function of testing time.

The equation mean value function for leading faults is as follows:

$$\frac{dm_1(t)}{dt} = bt^k \times [a_1 - m_1(t)] \tag{14.9}$$

We have the following differential equation for dependent faults:

$$\frac{dm_2(t)}{dt} = ct^k \times [a_2 - m_2(t)] \times \frac{m_1(t-\phi(t))}{a_1} \tag{14.10}$$

Model 4: The debugging time lag functions discussed in Section 14.3.2 have been used, on the basis of which the following mean value functions have been derived.

If $\phi(t) = 0$, then solving Equations (14.9) and (14.10) at boundary conditions and by using Equation (14.3), we get the following model:

$$m(t) = a\left(1 - p\exp\left[-\frac{bt^{k+1}}{k+1}\right] - (1-p)\exp\left[\frac{c}{b}\left(1-\exp\left[-\frac{bt^{k+1}}{k+1}\right]\right) - \frac{t^{k+1}}{k+1}c\right]\right)$$

Model 5: If $\phi(t) = t - \left[t^{k+1} - \frac{k+1}{b}\log\left(1+\frac{bt^{k+1}}{k+1}\right)\right]^{\frac{1}{k+1}}$, then solving Equations (14.9) and (14.10) at boundary conditions and by using Equation (14.3), we get the following model:

$$m(t) = a\left(1 - p\exp\left[-\frac{bt^{k+1}}{k+1}\right] - (1-p)\right.$$

$$\left.\exp\left[\frac{2c}{b}\left(1-\exp\left[-\frac{bt^{k+1}}{k+1}\right]\right) - \frac{t^{k+1}}{k+1}c\left(1+\exp\left[-\frac{bt^{k+1}}{k+1}\right]\right)\right]\right)$$

Model 6: If $\phi(t) = t - \left[t^{k+1} - \frac{k+1}{b} \log\left(1 + \frac{bt^{k+1}}{k+1} + \frac{b^2}{2}\left(\frac{t^{k+1}}{k+1}\right)^2 \right) \right]^{\frac{1}{k+1}}$, then solving

Equations (14.9) and (14.10) and by using Equation (14.3), we get the following model:

$$
m(t) = a \left(\begin{array}{c} \left[1 - p\exp\left[-\frac{bt^{k+1}}{k+1}\right] - (1-p) \right] \\[4mm] \exp\left[\begin{array}{c} \frac{3c}{b}\left(1 - \left(1 + \frac{bt^{k+1}}{k+1}\right)\exp\left[-\frac{bt^{k+1}}{k+1}\right]\right) \\[4mm] -\frac{t^{k+1}}{k+1}c\left(1 - \left(1 - \frac{bt^{k+1}}{2(k+1)}\right)\exp\left[-\frac{bt^{k+1}}{k+1}\right]\right) \end{array} \right] \end{array} \right)
$$

Model 7: If $\phi(t) = t - \left[-\frac{1}{b}\log\left(\frac{(1+\beta)\exp\left(-\frac{bt^{k+1}}{k+1}\right)}{1+\beta\exp\left(-\frac{bt^{k+1}}{k+1}\right)} \right)^{k+1} \right]^{\frac{1}{k+1}}$, then solving

Equations (14.9) and (14.10) at boundary conditions and by using Equation (14.3), we get the following model:

$$
m(t) = a\left(1 - p\exp\left(-\frac{bt^{k+1}}{k+1}\right) - (1-p)\exp\left[-\frac{ct^{k+1}}{k+1}\right]\left[\frac{1+\beta}{1+\beta\exp\left[-\frac{bt^{k+1}}{k+1}\right]} \right]^{\frac{c(1+\beta)}{b\beta}} \right)
$$

14.4.5 Power Function–Based Fault Dependency Models by Considering Time Lag in Leading Faults [10]

In this section, some mathematical models have been proposed by considering the dependence of faults using a power function of testing time. In this case, it is assumed that during the removal of leading faults, the fault is first detected, isolated, and then removed. And in any case, we cannot neglect the debugging time lag for independent faults. In this section, we have used the debugging time lag functions defined in Section 14.3.2.

The leading faults removal depends upon the remaining leading faults in the software system. We consider the following differential equation to model the leading faults removal:

$$\frac{dm_1(t)}{dt} = bt^k \times [a_1 - m_1(t)] \tag{14.11}$$

Solving the preceding equation with the initial condition, that is, at $t = 0$, $m_1(0) = 0$, we have

$$m_1(t) = a_1\left(1 - \exp\left[-\frac{bt^{k+1}}{k+1}\right]\right) \tag{14.12}$$

Under the assumptions stated, we have the following differential equation for dependent fault removal:

$$\frac{dm_2(t)}{dt} = ct^k \times [a_2 - m_2(t)] \times \frac{m_1(t - \phi(t))}{a_1} \tag{14.13}$$

Solving the preceding equation with the initial condition at $t = 0$, $m_2(0) = 0$, and using Equation (14.3), and different debugging time lag functions discussed in Section 14.3.2, we obtain $m(t)$ for different cases which are discussed in the text that follows.

Different debugging time lag functions have been considered for both types of faults: leading and dependent.

Model 8: If $\phi(t) = 0$ in Equation (14.13), then by using Equation (14.3), we have the following mean function for total fault removal:

$$m(t) = a\left(1 - p\exp\left[-\frac{bt^{k+1}}{k+1}\right] - (1-p)\exp\left[\frac{c}{b}\left(1 - \exp\left[-\frac{bt^{k+1}}{k+1}\right]\right) - \frac{t^{k+1}}{k+1}c\right]\right)$$

Model 9: If $\phi(t) = t - \left[t^{k+1} - \frac{k+1}{b}\log\left(1 + \frac{bt^{k+1}}{k+1}\right)\right]^{\frac{1}{k+1}}$, Equation (14.13) becomes

$$\frac{dm_2(t)}{dt} = ct^k \times [a_2 - m_2(t)] \times \frac{a_1\left(1 - \left(1 + b\frac{t^{k+1}}{k+1}\right)\exp\left(-b\frac{t^{k+1}}{k+1}\right)\right)}{a_1}$$

Solving the preceding equation with the initial condition $t = 0$, $m_2(0) = 0$, and using Equation (14.3), we obtain the following:

$$m(t) = a\left(1 - p\left(1 + \frac{bt^{k+1}}{k+1}\right)\exp\left[-\frac{bt^{k+1}}{k+1}\right] - (1-p)\right.$$

$$\left.\exp\left[\frac{2c}{b}\left(1 - \exp\left[-\frac{bt^{k+1}}{k+1}\right]\right) - \frac{t^{k+1}}{k+1}c\left(1 + \exp\left[-\frac{bt^{k+1}}{k+1}\right]\right)\right]\right)$$

Model 10: If $\phi(t) = t - \left[t^{k+1} - \frac{k+1}{b}\log\left(1 + \frac{bt^{k+1}}{k+1} + \frac{b^2}{2}\left(\frac{t^{k+1}}{k+1}\right)^2\right)\right]^{\frac{1}{k+1}}$, Equation

(14.13) becomes

$$\frac{dm_2(t)}{dt} = ct^k \times \left[a_2 - m_2(t)\right] \times \frac{a_1\left(1 - \left(1 + b\frac{t^{k+1}}{k+1} + \frac{b^2}{2}\left(\frac{t^{k+1}}{k+1}\right)^2\right)\exp\left(-b\frac{t^{k+1}}{k+1}\right)\right)}{a_1}$$

Solving the preceding equation with the boundary condition $t = 0$, $m_2(0) = 0$, and using Equation (14.3), we obtain

$$m(t) = a\left(\begin{array}{l} 1 - p\left(1 + \frac{bt^{k+1}}{k+1} + \frac{b^2}{2}\left(\frac{t^{k+1}}{k+1}\right)^2\right)\exp\left[-\frac{bt^{k+1}}{k+1}\right] - (1-p) \\ \\ \exp\left[\begin{array}{l}\frac{3c}{b}\left(1 - \left(1 + \frac{bt^{k+1}}{k+1}\right)\exp\left[-\frac{bt^{k+1}}{k+1}\right]\right) \\ \\ -\frac{t^{k+1}}{k+1}c\left(1 - \left(1 - \frac{bt^{k+1}}{2(k+1)}\right)\exp\left[-\frac{bt^{k+1}}{k+1}\right]\right)\end{array}\right]\end{array}\right)$$

Model 11: If $\phi(t) = t - \left[-\frac{1}{b}\log\left(\frac{(1+\beta)\exp\left(-\frac{bt^{k+1}}{k+1}\right)}{1+\beta\exp\left(-\frac{bt^{k+1}}{k+1}\right)}\right)^{k+1}\right]^{\frac{1}{k+1}}$, Equation (14.13)

becomes $\dfrac{dm_2(t)}{dt} = ct^k \times \left[a_2 - m_2(t)\right] \times \dfrac{a_1\left(1 - \exp\left(-\frac{bt^{k+1}}{k+1}\right)\right)}{a_1\left(1 + \beta\exp\left(-\frac{bt^{k+1}}{k+1}\right)\right)}$

Solving the preceding equation with boundary condition $t = 0$, $m_2(0) = 0$, and using Equation (14.3), we obtain

$$m(t) = a \left(\begin{array}{c} 1 - p \dfrac{(1+\beta)\exp\left[-\dfrac{bt^{k+1}}{k+1}\right]}{1+\beta\exp\left[-\dfrac{bt^{k+1}}{k+1}\right]} - (1-p) \\[6ex] \exp\left[-\dfrac{ct^{k+1}}{k+1}\right]\left(\dfrac{1+\beta}{1+\beta\exp\left[-\dfrac{bt^{k+1}}{k+1}\right]}\right)^{\frac{c(1+\beta)}{b\beta}} \end{array} \right)$$

14.5 Conclusion

We have presented a framework of software reliability growth models by considering the fault dependency concept and various debugging time lag functions. We have taken two types of faults: leading and dependent. We have discussed the different cases: (1) fault dependency-based software reliability growth models by considering both leading faults and dependent faults where both types of faults assume debugging time lags, (2) fault dependency-based software reliability growth models by considering both leading faults and dependent faults where only dependent faults assume debugging time lags, (3) power function of testing time-based fault dependency-based software reliability growth models by considering both leading faults and dependent faults where only dependent faults assume debugging time lags, and (4) power function of testing time-based fault dependency-based software reliability growth models by considering both leading faults and dependent faults where both leading and dependent faults assume debugging time lags. The models discussed in the chapter will support software managers in improving the quality of the software products.

References

[1] Schneidewind, N.F. (1975) Analysis of Error Process in Computer Software, Sigplan Notices, Vol. 10, No. 6, pp. 337–346.

[2] Schneidewind, N.F. (2002) An integrated failure detection and fault correction model, In: *proceedings of 18th International Conference on Software Maintenance (ICSM 2002)*, Montreal, Quebec, Canada, Oct 2002, pp. 238–241.

[3] Schneidewind, N.F. (2001) Modelling the fault correction process, In: *Proceeding of 12th International Symposium on Software Reliability Engineering*, ISSRE, pp. 185–190.

[4] Xie, M. and Zhao, M. (1992) The Schneidewind software reliability model revisited, In *Proceedings of the 3rd IEEE International Symposium on Software Reliability Engineering* (ISSRE'92), Research Triangle Park, North Carolina, USA, Oct 1992, pp. 184–192.

[5] Kapur, P.K. and Younes, S. (1995) Software reliability growth model with error dependency, *Microelectronics and Reliability*, Vol. 35, No. 2, pp. 273–278.

[6] Kapur, P.K., Bardhan, A.K. and Shatnawi, O. (2001) Software reliability growth model with fault dependency using lag function, *International Conference on Quality, Reliability and Control*, IIT Mumbai, Dec 26–28, pp. R53-1-7.

[7] Huang, C.Y. and Lin, C.T. (2006) Software reliability analysis by considering fault dependency and debugging time lag, *IEEE Transaction on Reliability*, Vol. 55, No. 3, pp. 436–449.

[8] Kapur, P.K., Singh, V.B. and Yadav, K. (2007) Software reliability growth model incorporating fault dependency concept using a power function of testing time, In: Kapur, PK and Verma, AK (Eds.), *Quality Reliability and Infocom Technology*, MacMillan India Ltd., India, pp. 587–595.

[9] Kapur, P.K., Singh, V.B. and Anand, S. (2007) Fault dependency based software reliability growth modeling with debugging time lag functions, communications in dependability and quality management, *An International Journal, Serbia*, Vol. 10, No. 3, pp. 46–68.

[10] Singh, V.B., Yadav, K., Kapur, R. and Yadavalli, V.S.S. (2007) Considering fault dependency concept with debugging time lag in software reliability growth modeling using a power function of testing time, *International Journal of Automation and Computing*, Vol. 4, No. 4, pp. 359–368.

[11] Chatterjee, S. and Shukla, A. (2016) Change point -based software reliability model under imperfect debugging with revised concept of fault dependency, *Proc IMechE Part O:Journal of Risk and Reliability*, Vol. 230, No. 6, pp. 1–19.

[12] Goel, A.L. and Okumoto, K. (1979) Time dependent error detection rate model for software reliability and other performance measures, *IEEE Transactions on Reliability*, Vol. R-28, No. 3, pp. 206–211.

[13] Yamada, S., Ohba, M. and Osaki, S. (1983) S-shaped reliability growth modeling for software error detection, *IEEE Transaction on Reliability*, Vol. R-32, No. 5, pp. 475–484.

[14] Kapur, P.K., Bardhan, A.K. and Kumar, S. (2000) On categorization of errors in a software, *International Journal of Management and System*, Vol. 16, No. 1, pp. 37–38.

[15] Ohba, M. (1984) Inflection S-shaped software reliability growth model, Lecture notes In: Osaki, S. and Hotoyama, Y. (Eds.), *Economics and Mathematical Systems*, Springer Verlag, Berlin, Heidelberg, pp. 144–162.

[16] Kapur, P.K., Yadavalli, V.S.S. and Anu, G. (2006) Software reliability growth modeling using power function of testing time, *International Journal of Operations and Quantitative Management*, Vol. 12, No. 2, pp. 127–140.

[17] Kapur, P.K., Singh, V.B., Anand, S. and Yadavalli, V.S.S. (2008) Software reliability growth model with change-point and effort control using a power function of testing time, *International Journal of Production Research*, Vol. 46, No. 3, pp. 771–787.

Index

A

Acceptance testing, 209
"Accidents built-in design," 110
Additive quantile regression (AQR)
    models, 139, 141–142
AI, *see* Artificial intelligence (AI)
Algorithm GA–PSO-Co, 41–42
Alhazmi Malaya logistic (AML) model,
    125
AML model, *see* Alhazmi Malaya
    logistic (AML) model
AMSAA model, *see* Army Materials
    System Analysis Activity
    (AMSAA) model
Analysis/recommendation specification
    component, 167
Apache HTTP Server Project, 11–12
AQR models, *see* Additive quantile
    regression (AQR) models
Army Materials System Analysis
    Activity (AMSAA) model, 82
Artificial intelligence (AI), 1
    deep leaning approach, 6–9
    neural network approach, 4–6
Associated risks, 190
Attributes, 166, 177–178
AudioTrackData, 172

B

Bad data, 162
Bayes prediction intervals, 87–88
Behavioral similarity coefficient, 180
Behavioral view, 178–182
Beta distribution, 38
Big Data repository, 174, 175
Big data repository, 184
Big-M penalty function method, 40
Black-box testing, 216
"Black Swan" theory, 103–110

Branchand-bound programming
    method, 36
Branch coverage, 211
Brownian motion, 69
BTS, *see* Bug tracking system (BTS)
Bug tracking system (BTS), 2, 239
    of open source software, 3–4
Building blocks, 115

C

Calibrating defect density model,
    215–216
Capability maturity factor F_m, 214
Catastrophic effects, 20
Categorization of vulnerabilities,
    127–130
CB, *see* Content-based filtering (CB)
CCI, *see* Conditional confidence interval
    (CCI)
CCR splines, *see* Cyclic cubic regression
    (CCR) splines
CDF, *see* Cumulative distribution
    function (CDF)
CF, *see* Collaborative filtering (CF)
C-INCAMI
    framework, 165–169
    MIS, 169–172
Classifiers, 184
Cloud software, 9
Cobb–Douglas type function, 52
Code churn factor F_{cc}, 213, 215
Code execution, 123
Code reuse factor F_{ru}, 213, 215
Coding phase, 208
Collaborative filtering (CF), 114
Common vulnerability scoring system
    (CVSS), 122
ComplementaryData tag, 171
ComplementaryDatum class, 168
Computer-based automated system, 65

Conditional confidence interval (CCI), 81
Conditional coverage probability of unconditional confidence interval, 97
Conditional testing, 93–98
  for β, 88–89
Confidence interval, 84–85, 93–98
  based on future observations, 85–87
Constraint-handling technique, for constrained mixed-integer nonlinear problems, 40
Constriction coefficient–based particle swarm optimization, 44
Content-based filtering (CB), 114
Context properties, 166
Context specification, 166
Continued testing, 20
Continuous integration server, 4
Continuous rank probability score (CRPS), 144
Cost modeling, 22
Cost-reliability-optimal testing, 58
Counting process, 80–81
Countless optimization models, 22
Cross-site request forgery, 123
Cross-site scripting (XSS), 123
XSS, *see* Cross-site scripting (XSS)
CRPS, *see* Continuous rank probability score (CRPS)
CR splines, *see* Cubic regression (CR) splines
Cubic regression (CR) splines, 140
Cumulative distribution function (CDF), 150
CVSS, *see* Common vulnerability scoring system (CVSS)
Cyberattacks, 20
Cyclic cubic regression (CCR) splines, 140

D

Daily peak electricity demand (DPED), 141
DataCollectorName, 169
Data contents, for open source software, 3–4
Data-driven decision making, 162

Data gathering, 163
Data mining, 115
Data quality model, 162
Data Set-I, 72
DataSourceAdapter class, 169
DataSource class, 169
Data source property, 169
DE, *see* Differential evolution (DE)
Debugging time lags, 244
Decision maker, 161, 178, 183, 184, 185
Deep learning (DL), 1
Defect detectability, 212
Defuzzification of triangular fuzzy number, 38
Denial of Service (DoS), 123
Dependent faults, 238
Design phase, 208
Deterministic measure, 167
DeterministicMeasure class, 167
DeterministicValue tag, 171
Differential evolution (DE), 36
Directory traversal, 123
Discrete-time markov chain analysis, 157
DL, *see* Deep learning (DL)
DoS, *see* Denial of Service (DoS)
Douban datasets, 113
DPED, *see* Daily peak electricity demand (DPED)
dsAdapterID property, 170
DS-I
  comparison criteria for, 73
  goodness-of-fit curve for, 74
  parameter estimates for, 73
DS-II
  comparison criteria for, 73
  goodness-of-fit curve for, 74
DS-III
  comparison criteria for, 74
  goodness-of-fit curve for, 75
  parameter estimates for, 73
DS-IV
  comparison criteria for, 74
  goodness-of-fit curve for, 75
  parameter estimates for, 73
Duane model, 82
Duane plot, 82
Dynamic programming, 36

E

Electricity demand forecasting, 138
Elementary indicator, 167
Entity under analysis, 170, 180
EPLF, *see* Extreme peak load frequency
    (EPLF)
Erlang model, 67
Error
    dependency model, 242–243
    generation, growth pattern of, 71
    measures, 144–145
Evaluation design and implementation
    component, 167
Expected value of beta distribution, 38
Exploratory data analysis, 145
Exponential-based stochastic
    differential equation, 67
Exponential reliability function,
    156–157
Extreme peak electricity demand
    estimation of threshold, 150–154
    nonlinear detrending, 150
Extreme peak load frequency (EPLF),
    138

F

Failure count data, 30
Fault big data analysis, 1–14
Fault correction process, 239–242
Fault dependency, modeling and
    analysis of, 237–250
Fault patching, 21–22
Fault removal phenomenon, 65–70
FCMs, *see* Fuzzy cognitive maps (FCMs)
Federal State Statistics Service data, 104
Field testing phase of software, 25
File inclusion, 123
First interval $(0,\tau)$, 25
Fisher information, 85, 89, 91
Forecast combinations, 144–145
Forecasted extreme peak loads, 154
Forecasting electricity demand, 143–144
FreeBSD 4.0, 124
Fukushima Daiichi Nuclear Power Plant
    accident, 109
Full likelihood, estimates and
    confidence limits based on, 86

Functional testing, 216
Future reliability, 98–99
Fuzzy cognitive maps (FCMs), 114
Fuzzy number, 37
Fuzzy set, definitions, 37–38

G

GA, *see* Genetic algorithm (GA)
Gain information, 123
Gain privilege, 123
GAMs, *see* Generalized additive models
    (GAMs)
Gathering function, 184
Generalized additive models (GAMs),
    139–141
Generalized Pareto distribution (GPD)
    model, 151
Genetic algorithm (GA), 42–43
GeographicComplementaryData, 171
Global indicator, 167
GOCAME, 167
Goel–Okumoto model, 66
GPD model, *see* Generalized Pareto
    distribution (GPD) model
Group recommender systems, 115
Growth pattern of error generation, 71
Gumbel-type bivariate probability
    function, 57

H

Hadoop, 9
Hardware systems, 210
Homogeneous Poisson process (HPP),
    80
HPP, *see* Homogeneous Poisson process
    (HPP)
Hump-shaped curve, 125
Hybrid algorithm, 41
Hypertext transfer protocol response
    splitting, 123
Hypothesis tests for parameters, 88–93

I

IAEA, *see* International Atomic Energy
    Agency (IAEA)
Imperfect debugging, concept of, 72

Improved collaborative filtering technique, reliable recommender system using, 113–117
Inductive thinking, 107–108
Information-driven decision making, 163
Information filtering system, 113
Information technology, 162
In-house testing phase of software, 22, 25
In-memory filtering, 175
Input domain software reliability models, 218–219
Instantaneous discovery rate, trend in, 129
Integration testing, 208
Intensity function, 80
International Atomic Energy Agency (IAEA), 110
Internet software, security vulnerabilities in, 227–229
Irregular fluctuations, 69
ISO 25012, 162

J

Jester datasets, 113
Jira, 4

K

Kendall's correlation, 178
k-nearest neighbors algorithm (k-NN), 114
k-NN, *see* k-nearest neighbors algorithm (k-NN)

L

Large sample test for β, based on likelihood function, 89–90
Lasso variable selection, 139
Lessons learning, 106–107
LikelihoodDistribution class, 167
Likelihood estimation, 84–85
Linear vulnerability discovery (LVD) model, 125
Lingo, software package, 31

Littlewood–Verral Bayesian model, 222
LOLE, *see* Loss of load expectation (LOLE)
Loss of load expectation (LOLE), 138, 156
LVD model, *see* Linear vulnerability discovery (LVD) model

M

MAPLE software, 200
Maturity factor F_m, 213
MAUT, *see* Multi-attribute utility theory (MAUT)
Maximum likelihood estimates (MLE), 84
Mean square error (MSE), 31
Mean time between software failures (MTBF), 7
Mean time to failure (MTTF), 211
Mean Value Functions (MVFs), 24
Mean value of beta distribution, 38
Measurement
  design and implementation component, 166–167
  package, updating, 167–169
Measurement adapter, 184
Measurement and evaluation (M&E) process, 164
MeasurementItemSet tag, 170
Memory
  corruption, 123
  overflow, 123
M&E process, *see* Measurement and evaluation (M&E) process
MLE, *see* Maximum likelihood estimates (MLE)
Modeling software fault removal
  model formulation, 24–29
  notations, 23–24
  numerical example, 29–32
Modulated power law process, 82
MovieLens datasets, 113
MSE, *see* Mean square error (MSE)
MTBF, *see* Mean time between software failures (MTBF)
MTTF, *see* Mean time to failure (MTTF)
Multi-attribute utility theory (MAUT), 21

Multicomponent system reliability, 229–231
Multi-domain recommender systems, 115
Multiple-version programming, 230–231
Multi-release vulnerability model, 22
MVFs, *see* Mean Value Functions (MVFs)

N

National Vulnerability Database, 122
neural network (NN), 2
NHPP, *see* Nonhomogeneous Poisson process (NHPP)
NN, *see* neural network (NN)
Noncumulative discovery function, 126
Nonfunctional requirements specification, 166
Nonhomogeneous Poisson process (NHPP), 24, 80
Nonlinear ordinary least square, 29
Non-parametric method, 114
Nuisance parameter θ, test for β, 91–92
Numerical examples, 44–47, 59–62

O

OLS, *see* Ordinary Least Square (OLS)
OM, *see* Organizational memory (OM)
Open source software (OSS), 1, 2
    fault data on bug tracking system of, 3–4
    fault identification method, 15
OpenStack, 9
Operational forecasts, 148–150
Operational profile, 217
Optimal policies, 57–59
Optimal reliability allocation, 231–233
Optimal testing effort expending problems, 55–56
Optimization model, proposed, 196–199
Ordinary Least Square (OLS), 29
Organizational memory (OM) processing architecture, 174–175
OSS, *see* Open source software (OSS)

P

PAbMM, *see* Processing architecture based on measurement metadata (PAbMM)
Partially linear additive quantile regression (PLAQR) models, 139
Partial test data, 223–226
Particle Swarm Optimization (PSO), 43–44
Partition testing, 216
Patch, 21
P-distance algorithm, 115
Pedagogy, 20
Phase factor F_{ph}, 213
PictureData, 172
PlainTextData, 171
PLAQR models, *see* Partially linear additive quantile regression (PLAQR) models
PLP, *see* Power law process (PLP)
Point process characterization, 150–154
Power function
    based fault dependency models, 247–250
    of testing time–based fault dependency models, 246–247
Power law process (PLP), 81–87
Power of test, 92
Power systems reliability, 138
Predictive likelihood, 85, 86
Processing architecture based on measurement metadata (PAbMM), 172–174
Programming team factor F_{pt}, 213–214
PSO, *see* Particle Swarm Optimization (PSO)

Q

Quadratic form, test based on, 92–93
Quantile regression models, 139–141
Quantitative approach, *see also* Software reliability
Quantitativemeasure class, 167, 168
Quantitative measures, for software reliability assessment, 53–54

R

Randomness, 67
Random testing, 216
Real-life softwar failure data sets, 72–73
Real-time measurement/evaluation, as
    system reliability driver
    architecture based on measurement
        metadata, processing, 172–174
    C-INCAMI framework, 165–167
    limiting in-memory searching space
        related to organizational
        memory, 176–182
    measurement interchange schema
        based on C-INCAMI, 169–172
    measurement package, 167–169
    processing architecture
        based on measurement metadata,
            172–174
        organizational memory in,
            174–175
    weather radar of national institute
        of agricultural technology
        (Anguil), 169–172
Receiver operating characteristic curve,
    174
Recommender system, 113–116
Red Hat data, 124
Redmine, 4
Regression testing, 209
Release time problems, 20
Reliability
    analysis based on time series, 11–14
    analysis of power systems, 154–157
    assessment
        future, 98–99
        measures, 53–55
        for open-source software, 1–14
    growth process, 52
    optimization, 35
    of system elements, 108
Reliability redundancy allocation
        problem (RRAP), mathematical
        formulation of, 39–40
Reliable predictions of peak electricity
        demand, 137–157
Reliable recommender system, using
        improved collaborative
        filtering technique, 113–117

Repairable system modeling, using
        power law process
    bayes prediction intervals, 87–88
    conditional testing and confidence
        interval, 93–98
    counting process, 80–81
    future reliability assessment, 98–99
    hypothesis tests for parameters,
        88–93
    power law process, 81–87
Reserve margin, 154–156
Residual autocorrelation, modeling, 143
Revisiting error generation
    data analysis, 72–75
    methodology, 68–72
RMSE, *see* Root mean square error
        (RMSE)
Root mean square error (RMSE), 31
RRAP, *see* Reliability redundancy
        allocation problem (RRAP)

S

Safety, 104
Safety culture, 105–106
Sales model, proposed, 196
SARIMA, *see* Seasonal autoregressive
        integrated moving average
        (SARIMA)
SAS, *see* Statistical Analysis System
        (SAS)
Score function, test for β, based on, 90
SDE, *see* Stochastic Differential Equation
        (SDE); Stochastic differential
        equation (SDE)
SDLC, *see* Software Development Life
        Cycle (SDLC)
Seasonal autoregressive integrated
        moving average (SARIMA), 143
Second interval (τ, τ_p), 25–26
Security vulnerabilities, in internet
        software, 227–229
Shape parameter, 79, 81, 82, 88
Snapchat's third-party application, 20
Software
    defect density, 212–216
    failure
        component identification for, 9–11
        occurrence pattern, 55

failure data, 225
fault detection and correction process, 239–242
frameworks, 189–190
life cycle phases, 207–210
MAPLE, 200
reliability, 66
  growth models, 219–227, 237–250
  input domain, 218–219
  measurement/assessment, 51
  measures, 210–211
  models, 2
  tools for, 234
saturation period, 26
setup cost, 196–197
structure factor F_s, 214–215
testing, 190–191
  tools for, 234
vulnerabilities, categorization of, 121–134
Software Development Life Cycle (SDLC), 21
Software Product Quality Requirements and Evaluation (SQuaRe), 162
Software reliability growth modeling (SRGM), 20, 66, 206
Solaris 2.5.1, 124
Source code, 239
Spearman's correlation, 178
SQL injection, *see* Structured Query language (SQL) injection
SQuaRe, *see* Software Product Quality Requirements and Evaluation (SQuaRe)
SRGM, *see* Software reliability growth modeling (SRGM)
S-shaped models, 66
SSM, *see* Structural similarity matrix (SSM)
Statement coverage, 211
Statistical Analysis System (SAS), 29
Statistical computing, 184
Stochastic differential equation (SDE)-based software reliability growth models, *see also* Revisiting error generation
Structural similarity coefficient, 177
Structural similarity matrix (SSM), 177
Structural testing, 216

Structural view, 177–178
Structured Query language (SQL) injection, 123
Structure factor F_s, 213
Sufficient statistics, 83
Symantec Internet Security Threat Report, 121–122
System and data dependence characteristics, 162
System dependant characteristics, 162
System testing, 208–209

T

Test coverage, 211
Testing approaches, 216–218
Testing effort-dependent software reliability function, 54
Testing phase, 190, 208
TFN, *see* Triangular fuzzy number (TFN)
Third interval (τ_P, T_{LC}), 26–29
"Three Mile Island" (TMI-2) accident, 106
Time series
  analysis, 7–9
  reliability analysis, based on, 11–14
Time to next vulnerability (TTNV), 125
Total expected software cost, 199
Total time test plot, 82
TraceGroup class, 169
Transaction reliability, 210
Transport system safety indicator, dynamics of, 105
Triangular fuzzy number (TFN), 37
TROPICO R-1500, 72
TROPICO R-1500 switching systems, 29
TTNV, *see* Time to next vulnerability (TTNV)
Two-dimensional software reliability model
  basic assumptions, 53
  reliability assessment measures, 53–55

U

UCI, *see* Unconditional confidence interval (UCI)

Unconditional confidence interval (UCI)
    conditional coverage probability of,
        97
    for ϕ, 94
    for hydraulic systems of LHD
        machines, 98
Unconditional score test, 90
Unconstrained cost objective, 193
Unification scheme, 24
Unit testing, 208

V

VDMs, *see* Vulnerability discovery
        models (VDMs)
VideoData, 172
Vulnerability
    detection *see also* Modeling software
        fault removal
    in software
        modeling framework, 127–131
        numerical illustration, 132–133
        types, 123–124

Vulnerability discovery models (VDMs),
    124

W

Waterfall model, 207
Weather radar (WR) of national institute
        of agricultural technology
        (Anguil), 182–184
Weibull process, 80
White-box testing, 216
Wiener process, 69
Windows XP, 132
WinNT4, 124
Wireless sensor network (WSN), 169
WR, *see* Weather radar (WR)
WSN, *see* Wireless sensor network
        (WSN)

Y

YouTrack, 4